Most of our climate problems could be repaired by cleaning up the excess CO_2 in the air and so cooling things off. However, because of abrupt climate flips, the cleanup must be big, quick, and secure. Doubling all forests might satisfy the first two but it would be quite insecure—currently even rain forests are burning and rotting, releasing additional CO_2.

However, our escape route is not yet closed off. We can still do the equivalent of plowing under a cover crop, using perhaps one percent of the ocean surface for the next twenty years. A sustained bloom of algae is fertilized by pumping up seawater from the depths— whereupon another wind-driven pump flushes the surface water back into even deeper depths before its new biomass becomes CO_2 again. When the sunken biomass does decompose, its CO_2 is smeared out over 6,000 years. Such a slow return of excess CO_2 can be countered by forestry practices.

Putting current and past emissions back into secure storage would lower the global overheating, relieve deluge and drought, reverse ocean acidification, reverse half of sea level rise as the oceans cool, and reduce the chance of abrupt climate shifts. The plankton plantations could then be kept in readiness for cooling the planet in a methane emergency.

THE GREAT CO2 CLEANUP

Backing Out of the Danger Zone

WILLIAM H. CALVIN
University of Washington

ClimateBooks
@WilliamCalvin.org

William H. Calvin, Ph.D., is a professor at the University of
Washington in Seattle and the author of *Global Fever: How to
Treat Climate Change* (University of Chicago Press, 2008) and the
award-winning *A Brain for All Seasons: Human Evolution and
Abrupt Climate Change* (University of Chicago Press, 2002).
Website: WilliamCalvin.org.

Chapters

In memory of climate scientist
STEPHEN H. SCHNEIDER
(1945-2010)

Preface

In searching for an effective intervention for global warming, I found that existing ideas for recapturing the excess CO_2 from the air do not meet the test of being sufficiently big, quick, and secure. Here I sketch out a novel approach to ocean fertilization that appears to pass this triple test.

It mimics natural up- and downwelling processes using push-pull ocean pumps powered by the wind to pull sunken nutrients back up to fertilize the ocean surface—but then immediately pushes the new plankton production down to the slow-moving depths before it can revert to CO_2. Such a plankton plantation flushes not only the living biomass but also the hundred-fold larger amounts of dissolved organic carbon from feces and suspended debris.

Finally I discuss the looming methane problem and the emergency cooling of the planet that it will require. This too is urgent if we are to take effective action before being weakened by resource wars and economic collapse.

1
What to do about climate?

'What to do about climate' is not my favorite research subject. That would be how abrupt shifts in climate contributed to pumping up our brain size during the ice ages, the topic of my book *A Brief History of the Mind*.

But that's why, for almost thirty years, I've been closely following what climate scientists have to say about ancient climate changes. Paying attention also gave me a ringside seat for their analysis of global warming and the climate change mechanisms—and so what's likely to happen to our civilization unless we act swiftly to counter global warming.

We now have climate disease and our prospects are not good if we fail to quickly find an effective treatment.

I'm a great fan of the climate scientists—they've done a magnificent scientific job—but they didn't train to be climate doctors. No one did.

Though I'm not a physician, you can't hang around a medical school for fifty years without absorbing some of their way of thinking—and noticing its absence from our climate discussion.

For physicians, the window of opportunity for intervening successfully has become part of their training; they know how fleeting an opportunity can be, that too much emphasis on "certainty" can prove fatal by delaying effective treatment.

I'm afraid that for climate disease, the diagnosis, the prognosis, and the treatments don't fit together very well. The overall story is incoherent, however logical its parts and pieces.

The climate *diagnosis* represents excellent work by the climate scientists, who well deserve their 2007 Nobel Peace Prize.

But the *prognosis* and *treatment* aspects hang together about as well as the average nighttime dream, with its jumble of people, places, and years that don't hang together very well. They just don't make sense.

Much of what I had to say in the first book of this pair, *The Great Climate Leap: A Climate Surprise Is Like a Heart Attack,* pointed to a need to reframe the climate problem, to refocus on what's really important. Climate change is likely to be more sudden, with crises arriving sooner, than is usually thought.

So I'm going to present a "second opinion" about the climate *diagnosis*, the *prognosis* if untreated, and what *treatments* might actually fix the climate problem—rather than merely delaying civilization's collapse by a few decades.

Even though most proposed treatments are now insufficient—too little, too late—our situation turns out *not* to be hopeless. Things may still be retrievable at a cost that the economy can afford. But it has to be quick—as in the next few years.

Diagnosis

The earth is overheating. When we take carbon out of storage and put it into the air as CO_2 or methane, it traps additional heat.

The diagnosis of the climate problem is now quite certain: We know that fossil fuels and deforestation are the two big players, agriculture a minor player.

Farther down the list are the reductions in how much sunlight is bounced back out into space before it can heat the planet. The big example used to be golden waves of grain being replaced by blacktop roofs and parking lots. Now it's the loss of the nicely reflective Arctic sea ice in the summers when there is sunshine all through the night. It's one of those feedbacks that makes things automatically worse—and which were left out of past IPCC reports.

A lot of black soot floats around from incomplete burning—diesel trucks, low temperature cooking fires fueled by wood or dung. This has produced a brown cloud that hangs over parts of India and Southeast Asia. Soot absorbs sunlight and promptly re-radiates it as infrared heat, warming the air around the soot particle.

The sun's fluctuating brightness, often trotted out by climate skeptics, turns out to be about the smallest player in the overheating drama.

Quibbling about the diagnosis (using the scientists' own arguments from forty years ago—while ignoring their more recent answers) has been wasting the valuable time of the climate scientists.

Prognosis

Here again, for those aspects of future overheating that they have addressed, I think that the climate scientists have done an excellent job.

In half of the climate models[1], global average overheating is more than 2°C by 2048. But in the US, we get there by 2028—about the time that today's infants are teenagers.

Use less? Because most of the growth in emissions now comes from the developing countries burning their own fossil fuels to modernize with electricity and personal cars, we rich countries have little say in the matter.

But suppose the world succeeds. In the slow growth scenario, similar to what global emissions reduction might buy us, 2°C arrives by 2079 globally— but in the US, it arrives by 2037. *So drastic emissions reduction worldwide, however improbable that is, would still only buy the US about nine extra years.*

Yet milder forms of emissions reduction are all we ever talk about. You'd think that there was some sort of emerging taboo about discussing reality in polite company.

Note that *global* average temperature is averaged over day and night, over all four seasons, over poles and tropics, and over land and water. That approach, which comes from trying to account for the earth's energy budget, is like calorie counting for a diet. It really doesn't tell you most of what you need to know for the climate's prognosis.

For example, the European heat wave killed 70,000 people in the summer of 2003. But in the *global* average, it was not a particularly hot year–elsewhere it was somewhat cooler in other seasons of the year, holding the average down. But that didn't save those 70,000 people from being victims of climate change, triggered by the overheating of *previous* years.

Global average temperature–all you hear any talk about–is not going to be a good indicator of *when* our climate problem will become severe, causing mass extinctions of species and a collapse of civilization.

Also, the change in global *average* temperature hides the fact that the land is warming twice as fast as the oceans. That's one reason why the US and other large countries get a 2° hit so much sooner. And the Arctic is warming still more rapidly than land—the floating ice is melting and no longer reflecting most of the Arctic summer sunlight back out to space before it can heat things up.

Such uneven overheating also causes the onshore winds to be stronger. It's a response to the enhanced temperature contrast over the coastlines. Just like a hotter fireplace, the overheated land sucks in more air from that over the oceans.

But stronger winds can punch through old barriers, discovering shortcuts. Thus the ocean moisture may be delivered to places unaccustomed to so much rain. They get floods. The areas that formerly got this moisture get drought. So for the price of one detour, we get both deluge *and* drought.

Those hit-or-miss "detour droughts" are in addition to the more systematic drying that's occurring because the tropics are expanding from equatorial overheating. This mostly affects land near the temperate zone's boundaries with the subtropical deserts at 20-30 degrees from the equator.

What goes up must come down (with minor exceptions called spacecraft). The moist hot air that goes up to cause those heavy tropical rains comes back down in the hot subtropics like the Sahara, but missing the moisture. Little rain means there's no evaporative cooling of the ground, which makes desert latitudes hotter than the equator.

As the tropics expand, the dry Sahara is being pushed north. Its northward movement will deprive Mediterranean countries of their customary winter rainfall—which is why you've been hearing so much about droughts and fires in Mediterranean countries from Portugal to Greece.

The same thing is happening in other "Mediterranean climates" with their winter rain and summer sun—say, coastal southern California. Same thing for the Mediterranean climates on the coasts of Australia and Chile. And in southern Africa, the shifting Kalahari Desert is pushing that lovely garden strip near Cape Town south into the ocean.

Of course, since warmer air can hold more moisture, the world is actually getting wetter *on average*. But try telling that to the climate refugees fleeing the encroaching deserts—and encountering genocides.

Just remember that joke about the beginning student of statistics who drowned, wading in a pond that was, *on average*, only 3 feet deep. Averaging hides potentially fatal variations.

So while the cumulative effects of global overheating are indeed what's behind climate change, *civilization's* climate problem is not easily estimated from the average temperature trend.

How soon are we likely to get into big trouble? (Say, the partial collapse of civilization and its slippery slope that you may remember from reading Jared Diamond's book, *Collapse*.)

One thing that is seldom mentioned in the usual climate prognosis is that, at least five reasons, the time-to-disaster is likely to be far shorter than generally supposed. We overestimate the time remaining.

1. *Decade-scale climate fluctuations* will get to a tipping point sooner than the slow century-long trend itself (the first passage time effect in stochastic theory; my very first published paper back in 1965 applied first passage time math to threshold-crossing in nerve cells),

2. *abrupt climate shifts* (often from path changes in the "rivers" of ocean and air),

3. *intervening complications*,

4. *intentional omissions*, and perhaps some

5. *wishful thinking*.

The last three can be more quickly summarized than the first two.

Intervening complications. In physician-speak, a *serious* complication is the kind where the patient is more likely to die of the complication than from the original problem.

Fall and break a hip and you have about a 25% risk of dying from complications (infections, blood clots, and such). Another 50% of patients never return to their pre-injury level of activity. For morale-building, physicians often emphasize the success stories of the lucky 25% who fully recover. But success is not par for the course.

For climate, one of the most serious complications would be a big fire that suddenly put most of the Amazon rain forest's carbon up in the air as CO2, along with soot that heats up the air even more.

Or a complication could hinder recovery, as did the economic collapse in SE Asia that followed an oversized El Niño in 1997.

At least in medicine, *you cannot talk of prognosis without assessing the prospects of intervening complications.* We're not hearing much about this from the usual climate experts. What the climate scientists feel comfortable talking about–the part that their training has prepared them for–is not the whole story.

Omissions. In 1864, Lord Kelvin calculated the changes in the earth's temperature and concluded the earth could not possibly be as old as Charles Darwin said it was–it would have cooled off more if Darwin's time scale was correct. It wasn't until forty years later that Lord Kelvin learned about the heat produced by radioactive decay and how it kept heating the earth from within.

So, sometimes an omission is from ignorance. But here I want to talk of *intentional* omissions, usually for the climate and knock-on aspects that must be estimated from limited data. The search for quality turns out to produce an unwanted side effect.

IPCC climate scientists have been limiting themselves to discussing only the parts of the climate prognosis that they are really, *really* sure about. This tends to be the slow trends that are averaged over the whole globe. The IPCC models have until now simply omitted faster-acting effects based on thinner data–say, carbon cycle feedbacks that intensify CO_2 release in drought areas by cooking the parched soil.

The underestimates from intentional omissions have already been coming back to bite us.

1.   The 2007 estimates for sea level rise intentionally left out the contribution made by melting ice sheets (over the 2006 protests of many glaciologists), simply because others didn't feel sure about the numbers. They stuck to what they were sure about: thermal expansion of the oceans. Now we know that the 2007 sea-level-rise estimate was low by perhaps a factor of three.

2.   The last decade's emissions turned out to be as bad or worse than the high end of the 1990s emissions estimates–the *most pessimistic* ones weren't pessimistic enough.

3. The demise of Arctic sea ice has been happening much more quickly than what was estimated only a decade ago.

So a systematic understatement of our climate problems has resulted, mostly from this search for quality.

Understandable, you might say—but physicians learned long ago to be wary of letting "quality" and "certainty" arguments rule the day. They know there is a tradeoff because the associated delays can wind up harming the patient.

Then there is the state of the art for today's computer models. Climate scientists cannot yet talk with comparable certainty about jet stream detours and other sudden climate shifts.

What worries me most is that the usual climate discussion simply leaves out the sudden climate shifts that started happening in 1976. We have an observational track record that we seem to be ignoring when discussing the climate prognosis.

For example, world drought acreage suddenly doubled in 1982, stayed there until 1997—when it went to triple for eight years. They were likely triggered by the big El Niños of 1982 and 1997 but the excess droughts didn't go away when the year-long El Niño ended—things stayed "latched up." It's as if three *changes of state* happened—that history mattered, not just tracking the causes up and down each year.

Two more fast shifts can be seen in oceanographic data, and they are even more clearly changes of state. The ex-Gulf-Stream waters that go north past Ireland eventually sink into the ocean depths via giant whirlpools in both the Greenland Sea northeast of Greenland and in the Labrador Sea off the southwest of Greenland.

Normally both are flushing. But in 1978, the Greenland Sea whirlpools failed for a decade. [2] Then after nearly a decade of both working again, the Labrador Sea flushing failed in 1997,[3] resuming in 2007.[4] What if both were to fail at the same time?

So it doesn't take *steps* in temperature to get sudden steps in climate. They are occurring even with slow ramps in temperature. But climate modelers can't predict how or when such sudden shifts occur, so they leave it out of the public discussion.

Overall, I'd say that this is a case of the tail wagging the dog, when a small part's not-unreasonable scientific standards manage to set the style for addressing the bigger fuzzy part. (Or not addressing it.)

In general, the climate numbers are going to be low-ball estimates because of this intentional omission of many contributions. While filtering for quality (what climate scientists think they are doing) applies to both the pluses and the minuses, few of the things left out are likely to make things better; even the CO_2 fertilization of crops turned out to be small when tested in greenhouses with raised CO_2–and the positive effect on food production turned into a negative one when the overheating exceeded several degrees.

This asymmetric pruning of the tree of climate effects has meant that the IPCC numbers should be qualified, labeled "*At least this bad, but probably worse and probably sooner.*"

Yet climate experts seldom say so explicitly. I had to read a lot in this field before this omitted qualification dawned on me.

The whole emphasis on climate change "certainty" sounds really strange to medical-school ears. As we like to say, the physician who waits until dead certain before starting treatment is likely to wind up with a dead patient. Army generals probably have a similar aphorism for battlefield maneuvering. Waiting for certainty can prove fatal.

For climate, we have to distinguish between uncertainty in *what* and uncertainty in *when*. Most people are lumping them together. There is now little remaining uncertainty that something bad is already happening; it's no longer a matter of *if*. And, thanks to the butterfly effect in complex systems such as the atmospheric circulation, there will never be certainty regarding *when* and *where* a knockdown blow will hit us. It is almost guaranteed to be a surprise, however excellent the science.

Third, psychological factors may be minimizing even the low-ball version of the climate threat. I'm not speaking here of outright denial—few climate scientists have that problem—but something more subtle.

Wishful thinking about the time remaining. I don't know anything at all about how much wishful thinking there is among the climate scientists. But I can tell you how bad physicians are at estimating the time remaining.

For the terminal cancer patients that the physician is seeing as a consultant, only *one in three* lives as long as their estimate. They overestimate survival time; it's a systematic bias that continuing medical education might be able to recalibrate.

But if the physician knows the patient well, only *one in five* lives as long as what the physician writes in the chart. So wishful thinking is likely adding to the overestimate.

I've seen no sign that climate scientists are even aware of the psychological bias possibilities. Suppose we should assume that climate scientists, who know their patient well, are as bad as physicians?

So for multiple reasons, *when* we get into big trouble is likely to be substantially sooner than the mid-century years that you've heard about. What we do must now be quick-acting.

That has implications for the big question: *What to do about climate?* The prescribed long-term lifestyle changes–mostly variations on virtues such as "use less" and efficiency–will no longer suffice.

I have briefly covered the high-certainty climate *diagnosis* and the understated *prognosis*. What, however, about the possible *treatments* for climate disease? Will they be big enough and quick enough to do the job that now needs doing?

Treatments for Climate Disease

Unlike those excellent jobs on diagnosis and the aspects of prognosis that they have addressed, climate scientists haven't addressed treatment very much–and then only as a modeling parameter that represents the uncertainty in future emissions. The IPCC refuses to get into rescue programs beyond showing what happens for *generic* emission reduction scenarios provided to them by the economists–those emission reductions that cannot possibly fix our current climate problem in time.

In the official view, crafting climate fixes is someone else's business. Yet we all worry about how unlikely it is that legislatures will even understand the problem in time, given all of the obfuscation tactics of the powerful fossil fuel lobby. They mimic the earlier "blowing smoke" efforts by the tobacco industry that delayed action for forty years.

But doing nothing is not an option. We broke it. We've got to fix it—and quickly, just to save our own skins.

Even if we speedily reduced emissions, the earth's ecosystems would likely crash from the CO_2 accumulation, producing a mass extinction of terrestrial species.

And, because of ocean acidification, the crash there would start near the bottom of the food chain for aquatic species, percolating upward.

Some stewardship we've practiced, to get so much of life on earth painted into a corner, facing drastic population downsizing, if not extinction.

Prevention is now an outdated strategy. The need to reduce emissions was pretty obvious, even in 1950 when the *Saturday Evening Post* ran a prescient story on global warming.

Today, almost three generations later, we are finally getting started. But calculations[5] now suggest that "rapid deployment of low-emission energy systems can do little to diminish the climate impacts in the first half of this century." We cannot keep talking about emissions as if it was still 1950.

Alas, the climate fix is no longer a matter of just addressing the origins of the problem. By now, speaking *only* of emissions reduction has become the equivalent of a dentist telling you that your cavity could be fixed by brushing your teeth more often.

Yet prevention is all that seems to be on our climate agenda—just slowly reducing the use of fossil fuels.

This resembles locking the barn door after the horse is gone. It gives the *appearance* of action–without actually retrieving the horse.

And speaking of unfortunate misdirection, even the well-informed good guys talk as if emissions are *The Problem*. No.

The Climate Problem is the 40% excess of carbon dioxide already in the atmosphere, not its growth rate.

Emissions are merely the *rate* at which we make the Problem worse. It's like talking about miles per hour when what is important is miles traveled.

Excess CO_2 is what we have to reduce—and we cannot just wait a thousand years for natural processes to do 75% of the cleanup for us. There will be a mass extinction of species long before then, maybe even including humans.

Emissions reduction has now become a totally inadequate plan. Reducing emissions will still help slow things a few years but, by itself, this treatment will be "too little, too late."

Emissions reduction is like chemotherapy without removing the solid tumor first. In medicine, a therapy that is insufficient by itself, but which augments the power of another therapy, is called an adjuvant. That's all that emissions reduction is, by itself–a helper for a more effective treatment.

But we're not even talking very much about a more effective treatment, just promoting clean energy and electric cars.

Only a cleanup of excess CO2 *and* emissions reduction *in combination* will get us improvement and allow civilization's long-term survival.

Just who is it that decided that emissions reduction is the best we can do? Perhaps you see by now why I said that the prognosis and treatment for climate disease seems to be as incoherent as a nighttime dream, where things just don't hang together very well.

To judge what we should be doing, we need to know how big and how fast a cleanup must be before we design the project. The size-and-speed answers turn out to eliminate most of the CO2-scrubbing treatments that you might have heard about. But not all.

Size depends on speed so let's take speed first. Here's a summary of what I said in the first book of this pair, *The Great Climate Leap: A Climate Surprise is Like a Heart Attack.*

How fast? This depends on what might happen in the meantime. By my count, there have already been six abrupt climate shifts, global in scale, in the last 36 years– one every six years, on average, though there is nothing regular about them.

There are also about as many near-misses. So to be safer, make that one year in three. And that's not counting the big climate shifts that stay regional without going global.

So I'm going to take the next 20 years as the project time for cleaning up the CO2 excess. Ten years would be much better. Any plan that takes fifty years is going to be too late for most people.

And before you say that even 20 years is unrealistic, consider that during the next 20 years, we might be hit by another three abrupt shifts while we trying to fix the climate problem.

Should an abrupt shift so disorganize world agriculture that we start seeing a lot more resource wars and mass emigration, it could be the end of international cooperation and the beginning of a cruel downsizing of the human population via resource wars, famine, epidemic disease, and genocides–all four horsemen of the apocalypse.

How big? If 20 years is to be the duration of cleanup efforts, then how much annual sinking capacity must we aim for?

Note that it's not just the excess CO2 *currently* in the air that we have to clean up—even though that is exactly what is doing the harm. It turns out that the ocean will add more CO2 to the air during the cleanup.

The ocean surface layer has been buffering emissions (that's what is causing ocean acidification). For every eleven molecules of excess CO2 dissolved into the ocean from air bubbles in the crashing waves, ten of them are converted into carbonates and bicarbonates. But these are not permanent captures. It's not long-term storage, only an easy-in/easy-out buffer.

The reactions run backward as the air's CO2 concentration begins to drop, releasing some stored CO2 back into the air. That release from storage slows the cleanup by making the cleanup a bigger job than it first appears. (It's why we talk of "Total CO2" in buffered systems rather than partial pressures.)

So I have instead used the cumulative emissions since the Industrial Revolution—a total of 600 gigatonnes of carbon (10^{15}g of Carbon=GtC=PgC) as my target. We are at 350 GtC already, but we also have to counter the 250 GtC likely to be emitted during the project period.

Cleaning up 600 GtC at an average rate of 30 GtC each year ought to get us back to previously safe CO2 levels.

About a third of that would be going to cancelling out continuing emissions.

That 600 GtC we've taken out of storage happens to be about the same size as the amount of carbon currently stored in the forests of the world. So it's convenient to think of the cleanup as equivalent to quickly doubling the size of the world's forests, taking back the excess CO2 from the air and making wood out of it. That's the ballpark we must aim for.

Forests are not sufficiently secure. Whatever method we use to take the air's excess of CO2 out of circulation, it needs to be *big* (600 GtC) and it needs to be *quick* (less than 20 years). But there is a third essential: the method must also provide *secure* storage so that the stored carbon cannot get back into the air as CO2 or methane for thousands of years.

Thus doubling forests is not where we should place our bets. They can burn down too easily. Or simply rot away in a drought. And given that the climate forecast is for more heat waves, more drought, and higher winds, it is not going to be easy to protect even our current forests from fire and rot, let alone the new ones. Another 600 GtC could wind up in the air and double our problem.

The Amazon rainforests are already rotting from so many recent drought years, releasing additional CO2 into the air from storage. They have become a tinder box.

We once assumed that big fires couldn't happen in rain forests. Then they happened during the oversized El Niño of 1997. They have continued happening since then in the Amazon basin.

Our 600 GtC climate fix might need to be twice as big. And there is a second reason to aim higher in our cleanup plans.

The other large complication. Judging from the methane that has been bubbling up offshore of Siberia, we could suffer from a burp of methane from any continental shelf containing methane clathrates. It looks like ice, except that it burns.

It's found mostly in the North Pacific and the Arctic, the methane coming from anoxic organic sediments in ice age saltwater marshes that were then frozen during the ice age and subsequently inundated by sea-level rise as our warm period began 10,000 years ago. The methane was trapped in a matrix of frozen water, and this methane-water ice was then capped by new sediments as the shoreline receded. But little landslides in the new ocean floor can expose these shallowly buried layers, so that chunks of this ice pop to the surface and melt, releasing the methane into the air. And the increasing hydrostatic pressure on the sea floor from rising sea level can, via earthquakes, trigger landslides in drowned valleys that cause a worldwide increase in air temperature.

On the decadal time scale, methane can be a hundred times more potent at overheating than CO_2. A big release can cause the global temperature to spike. We would have to quickly remove the equivalent amount of CO_2 in order to offset the warming.

Even if my climate or ineffective-treatments analysis is off by a factor of two in size or speed, we still have to aim at a capture capacity of more than 30 GtC per year in order to have some reserve cooling capacity for a methane emergency.

Essentially, we can't just reduce emissions. Furthermore, the doubling-forests result also implies that we likely can't grow anything anywhere on land that will be quick enough and big enough for 30 GtC per year. And we can't wait for technological miracles that scrub so much CO_2.

Most of the ways out of this mess are now *too little too late*–and we didn't even notice the doors closing on us. Fortunately, we are not yet out of options. Just almost.

Here is a summary of what later chapters cover.

Plowing under the organic carbon crop. We can still tweak the natural carbon cycle in the oceans to remove more of the excess CO_2.

The basics: photosynthesis by algae removes CO_2 from the air, releasing the O_2 part while binding the carbon into organic molecules such as sugar. Yet most captured CO_2 goes right back into the air as living organisms of the ocean surface respire and then rot. That's the short loop of the carbon cycle, "atmospheric CO_2 back to atmospheric CO_2" rather than dust to dust, and in only a week or two.

But some of the larger fecal pellets and debris makes it down into the ocean depths. Most comes back up again– only a fraction of one percent becomes sediment–but until the upwelling, the sunken carbon is kept away from the air for a thousand years or more. That's the long loop– and it's more than 50,000 times slower than the short surface path.

It's the kind of long-term storage that might solve our climate problem: storing our excess 600 GtC down in that slow circulation through the ocean depths. We would be taking it off the fast track and rerouting it on the slow track.

We might sink the excess in 20 years but mixing during its journey ensures that it will come back up smeared out over thousands of years–and that's a 0.1 GtC per year reappearance rate which even low-tech forestry management could counter, unlike our 10 GtC/yr situation today.

The extra 600 GtC won't change things very much in the depths–less than 2 percent–because about 37,000 GtC is already floating around down there as total CO_2. Thanks to the fifty-fold dilution, the depths won't have the acidity problem that the surface waters currently have from the 30% increase in hydrogen ions floating around.

There are two possibilities for modifying the natural carbon cycle: *more photosynthesis* and/or *more sinking* of the new organic carbon compounds before they can revert to CO_2.

More photosynthesis by itself is plainly insufficient by now; to sink 600 GtC in 20 years, we would have to somehow fertilize all of the oceans enough to increase photosynthesis *four*-fold.

That's what it would take to get 600 GtC below the thermocline via settling out the heavier bits before they revert to CO_2. Sounds unlikely–another closed door, just like the one for emissions reduction.

Yet there has been little discussion of sinking the new organic carbon *before* it can revert to atmospheric CO_2– even though it's analogous to the farmer's fallow fields practice, back before synthetic fertilizers, of plowing under a cover crop to enrich the soil with atmospheric nitrogen.

My conclusion is that sinking more of the surface layer's suspended-in-solution organic carbon from feces and debris has become an essential part of the climate fix.

It contains a hundred times more carbon than that in the living organisms.

That is the summary of the analysis part of my argument in this pair of short books. At the heart of this second one will be a concrete example of how capturing and burying carbon might be done using paired push-pull pumps to augment natural up- and downwelling processes in the ocean.

There is some danger in presenting a concrete example of a cleanup project, as a number of scientists so far have forgotten my 90% analysis part because they were skeptical of the 10% concrete example. But I think it is important to show just how feasible a cleanup is, if we do it right away. Skepticism at this stage is appropriate; it will take a second Manhattan Project to get this right.

But, given the stakes, there is some danger in skepticism that is simply dismissive. This has already been happening with anything labeled geoengineering, and mostly because the best-known example looks like air pollution.

2
The Emergency CO2 Cleanup

"How did you go bankrupt?" one of the characters is asked in Ernest Hemingway's *The Sun Also Rises*.

"Two ways," he said. "Gradually and then suddenly."

Climate creep's gradual deterioration of our planet is bad enough. But it's those half-dozen climate leaps since 1976 (and a similar number of near misses) that tell us we have an emergency[1], that we are already in danger of a serious blow from which we might not be able to recover in time—with civilization collapsing and the human population crashing within, say, the next twenty years.

Abrupt climate shifts can be like heart attacks: some are minor, others catastrophic. There is no predicting which. And there is no predicting when. Only prevention works.

To back out of this danger zone for climate surprises, we must quickly recapture past CO_2 emissions from fossil fuels, currently totaling 1,300 gigatonnes—though usually we talk instead of the 350 gigatonnes of carbon in that excess CO_2. Note how different this is from proposals[2] to merely capture some of the smokestack CO_2 and store it. Cleaning up the excess CO_2 in the air should minimize the abrupt climate shifts by reducing the major intermediate cause: the land is heating up twice as fast as the ocean surface, setting up sudden circulation shifts.

As we try to remove the excess CO_2 over the next twenty years, another 250 GtC are likely be emitted, judging from the 3% annual growth in the use of fossil fuels. That's 600 GtC total that we need to take back[3]. It's comparable to four piles of coal as high as Mount Rainier. That's also about how much carbon that is held above ground in the remaining forests of the world. So the size of the twenty year cleanup job is equivalent to quickly doubling the forests of the world using fast-growing trees.

But in addition to being big and quick, our cleanup method must securely store that excess carbon so that it cannot get back into the air anytime soon. Given that the climate forecast is for drier droughts and stronger winds, forests will be very vulnerable to fire.

And if the fire doesn't get them, the rotting of dead trees will produce the CO_2 instead, taking a decade rather than a year. Even rain forests are currently burning and rotting, so doubling forests is not secure enough for our purpose. What is?

Chemically scrubbing the CO_2 from the air is expensive[4] and requires a lot of new electrical power from clean sources, not likely to arrive quickly enough. One cannot merely scale up what suffices on submarines.

That pretty much leaves us stuck with finding new ways of doing the 30 GtC/year cleanup with biology, where—like the new forests—CO_2 is captured by photosynthesis, the oxygen liberated, and the carbon incorporated into an organic carbon molecule such as sugar[5]. Then, to take it out of circulation, it must be buried.

Burying keeps organic carbon from decomposing or, if it does, keeps decomposition CO2 from returning to the air. (Fans of big words call this 'sequestration' but 'burial' will do here.)

One could, for example, bundle up all of the crop residue (half of the biomass grown each year is cornstalks, inedible leaves, shells, etc.) and sink the weighted bales to the ocean floor[6]. It will decompose there but it will be a thousand years before its CO2 can be carried back up to the ocean surface and into the air. Alas, this project, even when done on a global scale, will yield only a few percent of what we need to remove each year, which is 30 GtC.

And if crop residue represents half of the yearly agricultural biomass, this also tells you that additional land-based photosynthesis, competing for space and water with human uses, can't do the job in time. At best, it can only double the numbers for crop residue—and that's an order of magnitude short of what we need.

Miraculous new technology might come along but basically we must look to the oceans for the new photosynthesis and for the long-term storage of the CO2 thus captured.

Algal blooms are temporary increases in biological productivity when the ocean surface is provided with fertilizer containing missing nutrients such as nitrogen and phosphorus.

The bloom comes from a window of opportunity. It takes a week for the primary consumers of the new algae, the little animals called zooplankton, to catch up in sheer numbers, as they cannot reproduce every 11 hours like the algae can. Grazing reduces the algal population numbers over the following weeks, even though there is no shortage of nutrients.

Most of this bloom biomass turns right back into CO_2 via respiration and rotting. Aided by winds that stir the surface layers, this CO_2 escapes back into the air. Only one captured carbon in every four manages to settle into deeper waters, where it doesn't have a chance to mix with the atmosphere for another thousand years or so.

And only one sunken carbon atom in a thousand manages to become sediment on the ocean floor, taken out of circulation for a hundred million years. Half of the rest turns into CO_2 that rejoins the atmosphere when the deep water is first upwelled ("ventilated" in oceanspeak).

Recently the chemical oceanographers have learned that the other half of the dissolved organic carbon (called refractory DOC) is somehow protected from becoming CO_2 for about 6,000 years[7]. Despite being upwelled multiple times, it doesn't become atmospheric CO2 for a while. That's good news.

In addition to the heavier biomass (the larger fecal pellets and shells) that can settle into the depths before becoming CO2, there is an express route to the depths: surface waters are flushed via whirlpools into the depths of the Greenland Sea and the Labrador Sea.

Bulk flow is the difference between having a thin pill dissolve on your tongue and knocking back a fistful of pills using a gulp of water to carry them down. Whirlpools carry down the living biomass (from fish to bacteria) as well as the dissolved organic carbon (from feces and smaller cell debris—think of it as carbon soup).

The major attempt to fertilize new plankton growth has been sprinkling the ocean surface with iron filings[8]. In many areas of ocean, this promptly causes a bloom. But the amount of new growth that actually settles into the depths has been discouraging.

Fertilizing near the downwelling whirlpools sounded like a good way of sinking much more—but then I recalled that, since 1978, each of the two biggest downwelling sites had shut down for a decade. (Fortunately, they didn't both fail at the same time. See the near-miss section of *The Great Climate Leap*.) Cross off one more carbon storage scheme that isn't secure enough, though conceivably big enough and fast enough.

So let us consider floating windmills (or equivalent wave-powered pumps, powered by the wind at one remove). They could pump up the nutrients that accumulate in the ocean depths[9]. A second set of pumps could carry the enriched carbon soup down into the thousand-year depths, perhaps with some compressed air to prevent anoxia in the depths from the carbon soup's decomposition.

Those dissolved feces and such contain hundreds of times more organic carbon than does the living biomass—which is what makes downwelling surface water such a big deal compared to merely settling out the larger debris and fecal pellets.

In my preliminary estimates based on algal growth rates in algaculture[10], the plankton plantations would need to cover about 0.8% of the ocean surface, an area equal to that of the Caribbean. The push-pull pumps would likely be scattered around the outer continental shelves in prime fishing grounds belonging to well-developed countries that can afford it. Since the fish and fishermen will love a plankton plantation, there's a cognitive carrot: a yearly payoff in fish catch while growing the climate fix with its payoffs for everyone.

Just as farmers grow a nitrogen-fixing crop of legumes and then plow it under, we would be growing a carbon-fixing crop of plankton and then pumping it under.

This bears a resemblance to what an artificial kidney does for the patient in chronic renal failure: cleaning up the accumulating toxins. Indeed, dialysis is an excellent analogy for both climate creep and climate leap. Dialysis machines are also used to prevent heart attacks in aspirin overdose patients, who have a one-in-four chance of suffering a heart attack unless the excess aspirin is promptly removed from the blood stream.

Like the terminal kidney patients I saw in the 1960s, back before dialysis became more widely available, our civilization is starting to look like a goner—apt to trip into a vicious downsizing cycle of resource war, famine, pestilence, and genocide.

All it takes is a big climate leap that cripples agriculture long enough for big cities to crash. (And, before someone objects that big cities have survived big droughts before, note that this hypothetical famine is global in extent, so that food cannot be imported and emigration attempts are usually fatal.)

Push-pull pumping looks to be big, quick, and secure. It illustrates that there is still at least one escape route open to us that might cool things down, reverse ocean acidification, reverse the part of sea level rise that is due to the thermal expansion of the oceans, and likely back us out of the danger zone for climate leaps.

From concept to proof of principle, to demonstration project and then deployment often takes more than a decade—though with wartime priorities, World War II history shows that several years may suffice when multiple solutions are pursued simultaneously.

3
Flying without a Safety Factor

A long-term perspective may come from environmental and ethical concerns. Many people can achieve it without any specialty training. But scientists are often the only ones who can add up the numbers and attach them to a timeline, saying where water shortages will happen, what areas will become deserts, and how much sea level will surely rise.

For any given element needed by an ecosystem—say carbon, nitrogen, or phosphorus—we use an accounting procedure that looks like balancing a checking account's deposits and withdrawals, looking at whether the balance falls over time. If a mix of elements is needed to make photosynthesizing cells such as algae, and there is a scarcity issue for one component, then the situation is much like a production line shut down by a tardy supplier of a critical part. Anyone with a spreadsheet can now make a working model of such a construction process, tweaking the supply rate and the backup inventory to keep the production line going during brief supply disruptions—just as you would want immediately available cash on hand in case a sick relative needed a quick visit.

Climate modeling has focused on the averages and how temperatures ramp up. The state of the art is very good there and getting better, with a dozen groups around the world competing to do a good job of incorporating carbon cycle feedbacks—such as when melting the Arctic tundra adds additional CO2 and methane to the air, making it warmer still.

Note that warming the Arctic causes a chain reaction. Indeed, the process itself is much like the one for nuclear fission, where a uranium nucleus captures a slow neutron, becomes unstable enough to split into two lighter elements, and throws off two free neutrons to continue the mischief nearby (usually the packing density isn't high enough to keep the chain going). In the Arctic, we see excess CO2 creating even more excess CO2.

There are similar self-fulfilling contingencies such as rain forests burning down. The excess CO2 has already caused enough overheating to create additional CO2 via burning and rotting. (However "natural" this heating bonus, it is triggered by anthropogenic overheating.) Should chance create a long-lasting drought, half of the Amazon biomass' carbon could end up as atmospheric CO2 within a few years. It's like a fatal heart attack occurring before gradual heart failure can cause death.

Because of such things happening along the way, a temperature ramp rising to a tipping point seriously overestimates the time remaining. Brief peaks usually do the damage, long before changes in the average get around to it—as in heat waves or when the inundation from mean sea level rise is preceded by that from stronger storm surges.

Yet even the slowly escalating aspect of the threat has been repeatedly obscured by the way we talk at several removes from the core problem. Over the years, our CO_2 action plans have shifted focus in a way that has diverted us from the excess 240 GtC of CO_2 currently in the air (what actually drives ocean acidity and overheating). Instead we merely talk about slowing the rate of increasing the problem.

Our excess CO_2 problem somehow became 'stabilizing' CO_2 concentration—though a high (but unchanging) CO_2 concentration still results in an unstable climate that spins off surprises.

A commonly-heard opinion is that we should "Let the climate heal itself after cutting emissions." Framing our climate response solely in terms of gradually reducing emissions has caused a tendency—even among climate scientists—to treat carbon cleanup as a mere contingency plan should emission reductions fail ("Techniques for extracting atmospheric CO_2 ... might eventually prove necessary") rather than as a parallel approach of great value.

Many now think that taking the growth out of emissions is an adequate goal for responding to climate change. Or that achieving zero emissions sometime later this century will do the job. Either would be a significant achievement but neither would get rid of the excess CO_2 quickly enough, or in a way which avoids dangerous acidification of the surface ocean.

Our current climate response is mostly a re-emphasis of the old virtues, such as "use less." They do not constitute adequate treatment for the climate disease we already have. As when chemotherapy supplements tumor surgery, emissions reduction is merely an adjuvant (valuable to supplement another therapy but insufficient by itself). Our climate response must focus on forestalling catastrophic developments, so that old virtues will have time to work and innovations will have time to prosper.

Emission reductions will remain important for climate action, as they hasten the day and lower the needed pumping capacity. If we instantly stopped burning coal and forests, it would eliminate 100 GtC of the 250 GtC emissions estimated for the next 20 years, reducing the size of the needed fix from 600 to 500 GtC.

We need a safety factor proportional to the hazard: if we can overbuild stadiums by a factor of four in strength, surely preventing a collapse of civilization warrants an even larger safety factor. For such reasons we must front-load our climate response, much as a course of antibiotics may include a double dose the first day.

4
Sketching Out a Climate Fix

By 2003, even the holdouts among the climate scientists were becoming convinced that something serious was afoot. It was no longer merely a matter of a prediction that global warming would occur—and it wasn't just heat waves. The domino-effect consequences of overheating were showing up in too many places, such as the increasing numbers of forest fires, deluge, and drought. Glaciers in Greenland were surging.

And most of them realized that even if the data or mechanistic explanation turned out to be wrong for one indicator of overheating, that wasn't going to make the others go away. You couldn't just use uncertainties in thermometer readings to put concerns on the back burner again.

Yet most of public climate discussion remains at the level of what scientists were discussing fifty years ago. The questions policymakers are asking now were valid concerns back then but they have since been answered—though you'd never know this to hear the radio loudmouths, who ceaselessly flog the old uncertainties to gain attention to themselves (and thus their advertisers). They are making millions in advertising dollars by exploiting the ignorance of science in much of their audience.

While deforestation and fossil fuel emissions got us into trouble, *it does not follow that fixing them will get us out of trouble.* Emission reductions would only scale down the future CO2 additions—they do not subtract from the CO2 that has already accumulated. Though such reductions might suffice in the case of methane because so much naturally disappears in a decade, the natural drawdown mechanisms for CO2 take ten times longer[1], with a stubborn one-fourth still hanging around a thousand years later.

Because abrupt climate shifts have already begun surprising us, gradual overheating is no longer the correct focus for understanding the risk we now face, where fast tracks to disaster must be forestalled.

Can we back out of the danger zone for abrupt shifts, climate's version of a heart attack? In addition to a low-carbon diet, we need emergency repairs—an intervention analogous to dialysis, one that quickly cleans out the toxic substance. Judging from the rate at which climate surprises have been occurring, we may only have several decades remaining for an effective intervention. That is the time for completing the repairs, not that for getting started.

Indeed, our current plans for climate action look more like a buying-time therapy for late-stage cancer than like a physiological fix such as the dialysis for kidney failure that makes possible a near-normal patient. Here I ask what a "climate fix" might look like, one that relieves many of the ocean acidification and climate problems and makes possible a near-normal planet.

Considering the rate of abrupt shifts so far, it would be prudent to expect additional surprises soon. Any one of them might close off our escape route, so that we spiral up to catastrophic climate change. Most, however, could be avoided by taking the excess CO_2 out of circulation.

To back out of the danger zone for abrupt climate shifts, an appropriate goal would be to remove nearly all of the excess CO_2 over the next twenty years and put it back into long-term storage. About 30 GtC/yr must be taken out of circulation, one-third going to simply counteracting the out-of-control fossil fuel emissions that continue. And continue they will, especially in developing countries that will burn their own fossil fuels in order to modernize, their governments even less effective than the United States in implementing treaties.

The CO_2 cleanup would be an emergency repair, not a substitute for a low-carbon energy diet. Chemically scrubbing the atmosphere is unlikely to scale up fast enough. Though large enough, doubling forests[2] is not secure enough, given the trends in fire and drought. To avoid competing with the world's food production, most of any sequestered organic carbon must come from new biomass grown in new places.

Unless we quickly develop far more efficient methods for sequestering circulating carbon, we could see our escape route closed off.

Why the echo? No, you didn't lose track of what page you were on. There really are some sentences that repeat points made in earlier chapters. While some repeats are a teaching strategy, in this book there is an additional reason.

I am trying to craft each chapter so that, in addition to carrying along the main story through the book, any one chapter is sufficiently self-contained to stand alone as an op-ed or feature article, or to be assigned as reading for a course. Or so that you can email a chapter to someone without having to explain what was said in earlier chapters.

PDFs of individual chapters, with citations appended, can be downloaded from _WilliamCalvin.org/media_.

As the copyright page notes, any single chapter may be freely reproduced. No need to ask first, though a brief notification or clipping sent to _WCalvin@UW.edu_ would be appreciated.

5
Put the Genie Back in the Bottle?

Nearly all of the excess CO_2 is anthropogenic, an unintended byproduct of industrialization and such old technologies as cutting down trees and burning coal and oil. Unlike such threats to civilization as nuclear warfare, the climate crisis comes from very primitive, pickaxe-level technology.

Excess CO_2 is that above 280 parts per million in the air, the old maximum concentration for the last several million years of ice age fluctuations between 200 and 280 ppm. It is currently above 390 ppm and its increase since 1750 has been exponential.

The current 110 ppm excess puts us well into dangerous territory for climate surprises. At the 1976 sudden shifts, the excess was only 51 ppm, showing that a 350 ppm target[1], allowing for a 70 ppm anthropogenic excess, is not low enough. We hit 350 ppm back about 1988, well after the sudden shifts that began in 1976.

An appropriate goal for our emergency repair would be to "Put the genie back in the bottle"—to remove nearly all of the excess CO_2 and put it back into long-term storage—and to finish retiring this carbon debt within the next twenty years. It would be like drawing down a reservoir to keep a leaking dam from collapsing.

This emergency repair would need to recapture most of the 350 GtC fossil emissions between 1750 and 2009. On top of that, extrapolating the emissions trend since 2003 for two more decades suggests we should allow for capturing another 250 GtC to counteract out-of-control fossil fuel use.

If we could remove 30 Gt of circulating carbon each year, then in twenty years we could recapture an amount equal to all 600 GtC of the fossil fuels burned. Another 25 percent will be needed to offset deforestation. Additional amounts could offset such CO2 heating equivalents as methane leaks and brightness reductions.

Large, swift, and sure. If *how quick* is twenty years and *how big* is equivalent to doubling all land vegetation, then most carbon removal schemes[2] are too little, too late, or too insecure. Many such schemes can still be part of longer-term solutions. What we need in the near term, however, is a climate fix, an emergency repair to make the long-term solutions relevant.

Though there are schemes for scrubbing CO2 with proprietary chemicals[3], time is too short to rely on finding the power to run such a giant new industry. But it is not necessary to remove CO2 directly from the air to reduce its concentration there. Photosynthesis already removes about 210 GtC each year[4] to create organic carbon molecules such as sugar. For atmospheric CO2 to remain unchanged, we would expect 210 GtC/yr to be released as CO2 by cell respiration and decomposition (burning, rotting).

If we intercept some of the decomposition carbon before it reaches its usual atmospheric destination, the carbon cycle will become unbalanced, taking out more CO2 from the air than it returns to it. This draws down atmospheric CO2; it is the principle underlying most carbon removal schemes. But sea floor burial of crop residue[5] and sewage[6] would amount to less than 0.8 GtC/yr, not even able to counter the 2 GtC annually produced by continuing deforestation and quite inadequate for hiding 30 GtC/yr.

Even if the land and water could be found[7] to double forests, the climate forecast is for hotter summers, more droughts, and stronger winds. In some years, Amazonia already releases more CO2 than it absorbs. Planting trees simply does not qualify as a climate fix; they are insufficiently secure against fire and rot.

To avoid competing with the world's food production, most of any sequestered organic carbon must come from new biomass grown in new places—for example, sinking additional cell debris into ocean depths via fertilizing the algae[8].

In order to understand my suggestion that we sink the carbon in surface waters beneath the thermocline via bulk flow[9], we must first examine how 140 GtC worth of excess CO2 has already been absorbed by the oceans and how little of it is in thousand-year storage.

Figure 1

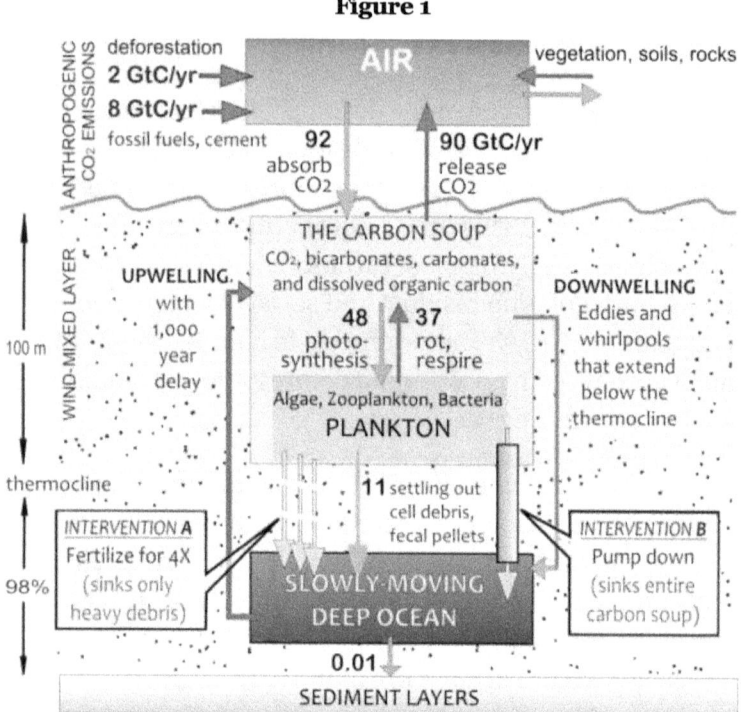

The Organic Carbon Soup

The ocean's carbon budget and how to keep 30 GtC/yr from becoming atmospheric CO2. Iron fertilization experiments have been aimed at settling more cell debris into the depths (*Intervention A*) but 30 GtC/yr would require 4X the natural settling worldwide, as would achieving fertilization by pumping nutrients up to the surface. Pumping down (*Intervention B*) can be done at many sites and sinks the entire "organic carbon soup" of surface waters—unlike *A*, where only the larger particulate matter settles quickly enough for its pending CO2 to be sequestered. (Source for fluxes: Houghton, 2007)

6

The CO2 Cache and the Conveyor Belt

Wind-driven waves serve to subdivide most areas of the
ocean into a wind-mixed surface layer and the slowly-moving
ocean depths. Excess CO2 in the air dissolves in the surface
layer as waves bury air, forming bubbles with much more
surface area. This creates a short-term easy-in-easy-out CO2
cache on top plus an out-of-contact storage loop through the
depths.

Any CO2 produced by respiration and decomposition in
the mixed layer will make it into the air. The bottom of the
wind-mixed surface layer— about 100 meters down in open
ocean but more like 30 m over a continental shelf)—is
marked by a decline in temperature. This "thermocline"
depth deepens as winter winds stir things more vigorously,
bringing up some of the nutrients that sank. That means
that, after the thermocline returns to 100 meters, the surface
layers are richer in nutrients for a while.

About 92 GtC is annually absorbed[1,4] into this wind-
mixed layer as dissolved CO2, with about 90 GtC being
released into the air. Most of the 2 GtC net gain in dissolved
CO2 is then buffered as bicarbonate (together, called total
CO2). Any attempt to draw down atmospheric CO2 by, say,
planting more trees will be slowed by the release of the
bicarbonate cache's excess (that's why I speak of recapturing

past emissions rather than just countering the excess CO2 in the air).

From this 92 GtC annual supply of CO2, 48 GtC is captured by algal photosynthesis to make organic carbon molecules such as sugar. Respiration and decomposition, however, soon makes CO2 out of 37 GtC of it. About 11 GtC is heavy enough to settle below the thermocline before decomposing. Once in the depths, the decomposition CO2 cannot easily reach the atmosphere. Most efforts to increase ocean sequestration, such as iron fertilization, have focused on settling additional dead biomass into the slowly circulating depths[2].

Algae are quickly grazed by zooplankton, which then respire and rot. The larger fecal pellets and the hard parts of plankton may sink through the thermocline before decomposing but smaller fecal pellets dissolve in the mixed layer. Dissolved organic carbon can survive in the mixed layer for a month[3] before bacteria consume it and, via their respiration, turn it back into CO2.

In addition to settling out 11 GtC/yr of debris and fecal pellets, there is that carried down in whirlpools. That biomass (algae, zooplankton, bacteria, exudates, debris, feces, and dissolved organic carbon) is hundreds of times larger than what succeeds in settling out. But this bulk flow into the depths only happens in a few places: small amounts in the Mediterranean Sea, more near the shores of Antarctica, most offshore of Greenland.

Because 200-km eddies and 20-km whirlpools carry surface waters far below the local thermocline, much of the anthropogenic CO2 sunk thus far into millennial depths is in the North Atlantic Ocean.[4] The "conveyor belt" that carries deep waters south has not been strong enough to keep it from accumulating in the North Atlantic.

Figure 2

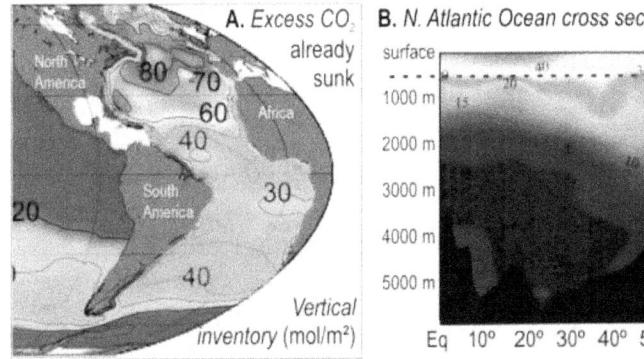

Excess CO2 in the oceans

A. In terms of water column inventory[5], the highest values are in the North Atlantic Ocean. Adapted from Sabine et al (2004).

B. The cross-section[6] for concentrations shows it is not due to mixed layer differences but to the amount sunk below 500 m (dashed line) by vertical convection. Adapted from Feely et al (2001).

While we usually think of the sunken debris as settling out on the ocean floor to become limestone or coal, nearly all circulates back up to the surface. A mere 0.02% stays behind as sediments. Most becomes dissolved "total CO2" in the slowly moving deep ocean.

Much of it resurfaces in the high southern latitudes, pumped up by the strong westerlies that circle the globe, pushing surface waters aside. Estimates for the delay until the flushed surface water resurfaces range from 400 to 1,600 years[7]. Excess carbon sunk to just below the winter thermocline is likely to resurface sooner. It may take a sinking depth of more than 1,000 m to achieve millennial-scale storage.

In the cold depths, about half of the new dissolved organic carbon from the upper ocean is promptly converted into total CO_2. But it has recently been shown that the rest has a 6,000 year residence time[8]. Since the reason for this postponed oxidation into CO_2 is not yet clear, one cannot say that half of the carbon debt, if sunk within twenty years, would also stay out of the atmospheric circulation for an extra 5,000 years. But it seems a good bet, one we should take.

If the 600 Gt carbon debt were sunk, this range of resurfacing delays would smear it out over more than 6,000 years, with release averaging 0.1 GtC/yr. This annuity-like pattern (an early lump-sum deposit with a delayed and distributed payout) can avoid the sharp peaks of temperature and acidity that cause the most harm. Surfacing CO_2 at 0.1 GtC/yr is also well within the range that even forestry management could counteract.

Our emergency repair, then, involves both efficiently creating new ocean biomass and efficiently sinking it into thousand-year depths before it reverts to CO_2.

In the Greenland Sea, the limiting nutrient is nitrate; winter deepening of the wind-mixed layer brings some to the surface. However, spring algal blooms can exhaust this new supply within several months. Summer algal population is then limited by what nitrate can be quickly recycled within the mixed layer.

Dust storms often fertilize offshore blooms; the iron fertilization experiments that they inspired do increase ocean productivity[9] but often do not settle out large quantities. Blooms are also triggered when a strong wind pushes around surface waters, thereby upwelling nutrients. This led to more recent suggestions[10] that vertical "ocean pipes" with a mechanical pump could similarly fertilize algal production with locally relevant nutrients, a continuous version of what winter winds bring up.

Fertilization alone can prove quite inefficient at sequestering carbon. To sink the needed 30 GtC/yr via settling out debris, we would need to quadruple primary production in all of the global oceans.

Clearly, boosting ocean productivity is not, by itself, the way to put the CO2 genie back in the bottle. Unless we quickly develop a far more efficient method for sequestering the excess circulating carbon, we could see our escape route closed off. That's how far we have already come in exceeding a safety margin.

The example in the next chapter shows a cleanup scheme that might suffice. It's not a ready-to-roll proposal so much as a challenge to the real experts, something to improve or replace in a solution space constrained by the need to be Big (600 GtC), Quick (20 yr), and Secure (for 1,000 yr). It shows a major recovery is still possible.

7
Plowing Under a Carbon-fixing Crop

To avoid competing with the world's food production and supplies of fresh water, most sequestered carbon must come from new biomass grown in new places. Here I explore how paired ocean pumps might uplift nutrients and then sink the new organic carbon back into the ocean depths.

Instead of sinking only the debris that is heavy enough to settle out, as in iron fertilization, we would be using bulk flow to sink the entire organic carbon soup of the wind-mixed layer (organisms plus the hundred-fold larger amounts of dissolved organic carbon) before its carbon reverts to CO_2 and equilibrates with the atmosphere.

The CO_2 later produced in the depths by the sunken carbon soup will reach the surface 400–6,000 years later. Smearing it out over that period greatly reduces the damaging peaks in ocean acidification and global fever.

Without fertilization, there is about one gram of organic-bound carbon in a cubic meter of seawater[1] in the North Atlantic. The North Atlantic's meridional overturning circulation thus sinks about 0.6 GtC/yr of organic carbon. That suggests that local concentrations about fifty-fold larger might sink our needed 30 GtC/yr—were we to rely only on the whirlpools.

But this too fails the *secure-enough* test. Each of the two major sinking sites has failed for a decade, just since 1978. Thus we need an estimate for plankton plantations that do their own sinking of surface waters.

If we fertilize via pumping up and sink nearby via bulk flow (a push-pull pump), we are essentially burying a carbon-fixing crop, much as farmers plow under a nitrogen-fixing cover crop of legumes to fertilize the soil. Instead of sinking only the debris that is heavy enough, we would be sinking the entire organic carbon soup of the wind-mixed layer.

Algaculture minimizes respiration CO_2 from higher up the food chain and so allows a preliminary estimate of the size of our undertaking. Suppose that a midrange 50 g (as dry weight) of algae can be grown each day under a square meter of sunlit surface, and that half is carbon. Thus it takes about $1 \times 10^{-4} \, m^2$ to grow 1 gC each year. To produce our 30×10^{15} gC/yr drawdown would require $30 \times 10^{11} \, m^2$ (0.8% of the ocean surface, about the size of the Caribbean).

But because we pump the surface waters down, not dried algae, we would also be sinking the entire organic carbon soup of the wind-mixed surface layer: the carbon in living cells plus the hundred-fold larger amounts in the surface DOC. Thus the plankton plantations might require only $30 \times 10^9 \, m^2$ (closer to the size of Lake Michigan).

Figure 3

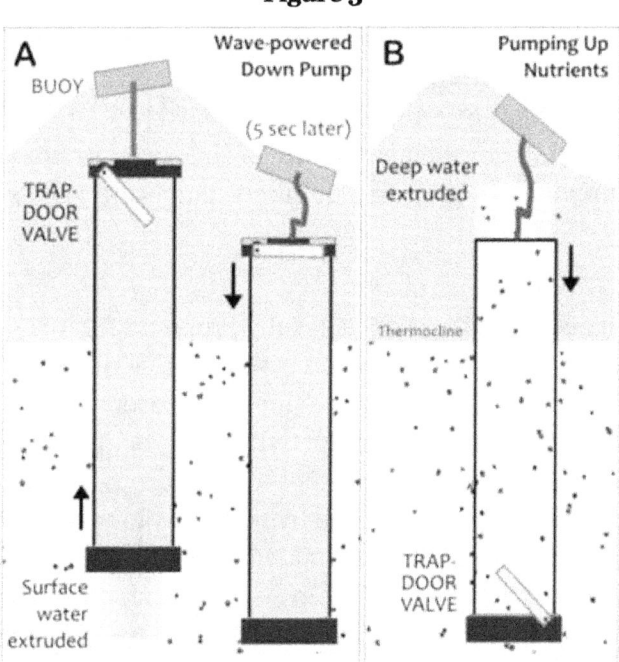

Push-Pull Pumps

A. An inexpensive wave-powered pump can push the surface's organic carbon soup below the thermocline. After a design by Philip Kithil, atmocean.com.

B. By reversing the trap-door valve, nutrient-laden cold water can be pulled to the surface. Schematics omit streamlining and ballast. For further discussion, see William H. Calvin (2008) *Global Fever: How to Treat Climate Change*. London and Chicago: University of Chicago Press.

The space requirement will be more because downpumps will not capture all of the new plankton; it might be less because the relevant algaculture focuses on oil-containing algal species and on harvesting a biofuel crop, not on plowing under the local species as quickly as possible. The ocean pipe spacing, and the volume pumped down, will depend on the outflow needed to optimize the organic carbon production[2]. Only field trials are likely to provide a better estimate for the needed size of sink-on-the-spot plankton plantations, pump numbers, and project costs.

Though ocean fertilization is usually proposed for low productivity regions where iron is the limiting nutrient, another strategy is to boost the shoulder seasons in regions of seasonally high ocean productivity. For example, ocean primary productivity northeast of Iceland drops to half by June as the nutrients upwelled by winter winds are depleted. Continuing production then depends on recycling nutrients within the wind-mixed layer. However, to the southwest of Iceland, productivity stays high all summer.

Because not all of the new plankton will be successfully captured and sunk, fertilization will stimulate the marine food chain locally. Most major fisheries have declined in recent decades and, even where sustainable harvesting is practiced, it still results in fish biomass 73% below natural levels[3]. At least for fish of harvestable size, there is niche space going unused.

Locating the new plankton plantations over the outer continental shelves is more likely to supply a complete niche for many fish species, whereas deep-water plantations will lack variety. (The main commercial catch in deep water is tuna.) Also, down-pumping near the shelf edge would deposit the organic carbon in the bottom's offshore "undertow" stream, carrying it over the cliff onto the Continental Slope into deeper ocean.

Note that pumps would be tethered to the bottom so that the ocean currents are always creating a plume downstream: a plume of fertilizer near the surface and a second plume of carbon soup in the depths. (Pumping up from a different depth than pumping down will prevent the interaction that characterizes the oceanographers' box models.) While the water might come back around in a thousand years, the plumes for the clean-up will only be about twenty years long and well diluted by that time.

Figure 4

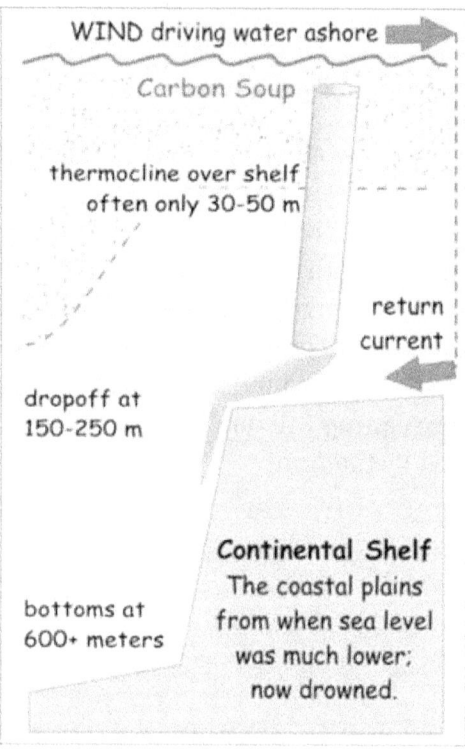

Augmenting the Continental Shelf Pump for carrying bottom water over the edge of the continental shelf and, if dense enough[4], into the ocean depths. Like the "carbon pump," it's only a metaphorical pump but it could be enhanced (green pipe) with a real pump, pushing carbon soup down to the bottom near the edge of the shelf. There it would be entrained by the return current from the wind-driven onshore current and carried down the continental slope into slowly circulating depths. Besides the shorter trip to the bottom, continental shelf siting also avoids international treaty complications, enabling fast action by a single nation.

Figure 5

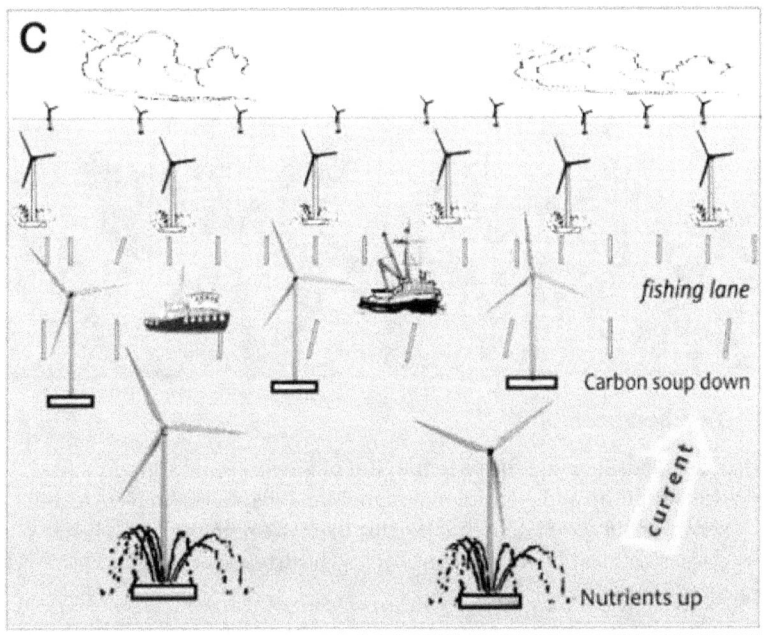

A plankton plantation

A design using windmill pumps, including a fishing lane free of anchor cables. From Calvin (2008) *Global Fever*.

Figure 6

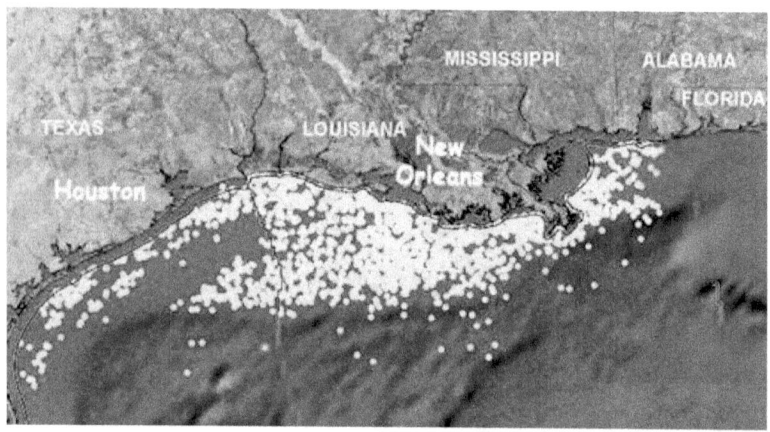

Test bed possibility

Existing drilling platforms in the Gulf of Mexico could support a field trial for pump and plantation design. Map (adapted from NOAA.gov) shows about 4,000 active platforms; there are perhaps 3,000 inactive ones. A similar opportunity exists in the North Sea.

8
Pro and Con

Here we have a candidate for removing 600 Gt of excess carbon from the air: the sink-on-the-spot plankton plantation that moves decomposition into the thousand-year depths. Push-pull pumping for fertilization and sequestration is relatively low-tech and merely augments natural up- and downwelling processes.

It has some unique advantages compared to current climate strategies:

- It is big, quick, and secure.

- It does not hinge on improbable self-denial by developing countries, treaties that take longer to negotiate than to ignore, or on finding a solution to the tragedy of the commons.

- It is impervious to drought and holdout governments.

- It does not compete for land, fresh water, fuel, or electricity.

- There is a "cognitive carrot," an immediate payoff every year (fish catch) while growing the climate fix (the 600 GtC emergency draw down).

It is against such advantages that we must judge the potential downsides. Concerns voiced thus far include:

1. *Could we get international agreement fast enough?* Continental shelves are within the Exclusive Economic Zone of the adjacent nation, which will have a strong interest in the restored fisheries. Shelves in the most productive latitudes belong to relatively wealthy countries. Their independent initiatives could quickly establish many plankton plantations without new treaties.

2. *Expensive?* If a pipe pump were to cost a thousand dollars in mass production, even a million of them would cost no more than the billion dollars spent in a single month on dog food in the U.S. But we cannot rely on any single method to work well, so we must budget on the thousand-fold-larger scale of what an unbudgeted bank bailout costs.

3. *Operating costs?* Wind and wave are free. Unlike fishing and farming, nothing about sink-on-the-spot plankton plantations will necessarily require handling, transport, or processing. Continuing costs would primarily be for maintenance and monitoring. Commercial fishing fees could provide significant income.

4. *Won't it pollute?* Perhaps not as proposed here, using local algae and nutrients in a vertical loop, but the usual considerations would apply should we want to introduce exotic or modified algal species to achieve even higher

rates of sinking potential CO_2. Toxic blooms are possible during productivity transitions.

5. *Won't anoxic "dead zones" form?* Shallow continental shelf sites might have to be thinly utilized because hypoxia would otherwise occur from the decomposetion of the downwelled carbon soup in a restricted volume. Fish kills will occur when anoxia initially develops and fish cannot find their way out of the anoxic zone in time. However, a maintained anoxic zone will mostly repel fish from entering the problematic plume. One advantage of windmill pumps is that compressed air could be bubbled into the Archimedes screw chamber so that sufficient oxygen would be available for the new CO_2 production in the depths.

6. *We don't know what will happen.* The novelty here is minimal, even less than for iron fertilization. Fertilizing and sinking surface waters merely mimics, albeit in new locations or new seasons, those natural processes seen on a large scale in winter mixing and in ocean up- and downwelling. There is also historical precedent. The 80 ppm drawdown of atmospheric CO_2 in the last four ice ages is thought to have occurred via enhanced surface productivity, aided by a major change in Antarctic offshore downwelling.

7. *Won't this just move the ocean acidification problem into the depths?* There are already massive downwells of surface waters with their 30% acidity increase. Since the depths are 98% of ocean volume, there is a fifty-fold dilution of this acidity. Sinking 600 GtC would add only

several percent to the deep-ocean inorganic carbon concentration—but that also means that we cannot keep up a 30 GtC/yr sinking rate for centuries without risking more serious impacts. After twenty years, we should be pumping only enough to offset any remaining out-of-control emissions. Another 2 GtC/yr may be needed to offset continuing deforestation. Additional amounts might offset such CO2 heating equivalents as methane leaks and the loss of surface brightness from ice melting.

8. *Isn't that a lot of water to flush through the depths?* The obvious test beds are the North Sea and Gulf of Mexico where thousands of existing drilling platforms could be used to support appended pipes and pumps. Without waiting for floating pumps, we could quickly test for impacts as well as efficient plantation layouts.

9. *What side effects might the pumps have on the North Atlantic's overturning circulation?* The two major downwelling sites are already wildly fluctuating, suggesting that a major component of current climate is quite unstable. Pumping up deep water in the right places might strengthen the whirlpools and help stabilize flushing's climate contribution. Such denser water should prime the surface waters to spiral down, thereby entraining less dense water for a net gain.

10. *Why not do it on land?* We could indeed do some of the algaculture in better-controlled conditions on coastal plains, with windmill pumps bringing up nutrient-laden water from just below the thermocline and then

returning it enriched with organic carbon soup to deeper ocean.

11. *Won't this take the pressure off the fossil fuel users?* This climate fix is an emergency repair analogous to drawing down a reservoir behind a leaking dam, not a model for how to do things in the future. The emissions-reduction agenda is still essential for life after the repair. In the same 20 years, we would need to stop the 3 GtC annual hit from burning coal and the 2 GtC from deforestation.

And three from the chemical oceanographers:

12. *Pumping up will just bring up water with higher CO2 than in the surface waters.* Vertical mixing, for example, causes the shallow southern part of the North Sea to export CO2 to the atmosphere in the summer[1]. In stratified North Atlantic waters, annual mean concentrations of inorganic carbon are about 2,170 μmol/kg in the 1500 m depths, 2,130 at 500 m, but with surface waters dropping as low as 1,970 in the northern North Sea[2]. A 40 μmol/kg difference means that upwelling brings up 0.48 gC m[-3] of inorganic carbon. Settling new biomass below the thermocline may yield less organic carbon, causing reasonable skepticism about up-only fertilization per se.

   In any event, this objection can be overcome by pairing down-pumps with up-pumps. Pumping down the same volume of *unfertilized* surface water would sink 1.0 gC m[-3] in the North Atlantic[3] as dissolved organic carbon (99%) and living biomass (1%), which will become CO_2 in the depths over 6,000 years.

The 0.48 gC m^{-3} "cost" (for a 40 μmol/kg depth-to-surface difference) is then subtracted from the gross 1.0 gC m^{-3} sunk, yielding a net sinking of 0.52 gC m^{-3} even with no fertilization. With fertilization creating a manyfold increase in dissolved organic carbon locally, the uplift cost is unlikely to overtake the potential CO_2 sunk.

13. *But nutrient depletion of surface water says that one cannot do something that big without an external source of essential nutrients.* Indeed, the 80 ppm CO_2 drawdown at the beginning of each of the last four glacial periods was initially attributed to continental dust storms fertilizing offshore waters[4].

There are several ways in which the usual box-model reasoning may not apply here. Plumes (the water moves but pumps are tethered) prevent the box-model-style mixing of up- and down-flows for millennia. Thus you continue to pump up deep water with pre-project levels of nutrients and inorganic carbon. This fertilizer spreads in a surface plume that widens out downstream of the tethered up-pump. Similarly a down-pump creates a plume in the depths that, because of differing pipe lengths, has little opportunity to feed back into the intake of the up pumps downstream. So instead of a box model and long-term thinking, we need a Heraclitus framing of the problem ("You cannot step twice into the same river") and a shorter-term focus to appreciate this opportunity for major carbon sequestration.

14. *Even so, aren't you going to run out of phosphate, what currently limits the global ocean productivity to a fraction of its capacity?* If not enough phosphate is recycled by the push-pull pumps, up-pump pipes could be sited to bring up bottom waters from the southern oceans that are currently rich in phosphate.

This brings up another big and secure (but not quick) carbon sequestration possibility—mimicking the ice age drawdown of 80 ppm CO_2. It was analyzed by Princeton's Robbie Toggweiler and colleagues[5].

Some essential background: deep water production near the Antarctic coastline works differently than do the deep whirlpools of the North Atlantic Ocean. In the Antarctic seas, there are holes in the pack ice where the downwelling takes place, particularly in the winter when fierce mistral-like downslope winds off the land keep a polynya open. (A polynya is an area of open water in otherwise closed pack ice.)

The intense evaporation at an opening leaves behind a lot of salt, creating exceptionally dense cold water that sinks and then slides down the continental slope to create Antarctic Bottom Water. This dense pool spreads north through the abyss as far as Chesapeake Bay.

Farther out from the Antarctic shores, westerly winds circle the continent without being slowed down via running into land. This pushes water ahead and to the sides, creating the Antarctic Circumpolar Current. The water pushed aside creates eddies and the upwelling of deep water: this is the major return path for the conveyor belt whose downwelling end is in the Greenland and Labrador Seas.

This southern ring of eddies is the major place where nutrients such as phosphate are moved from the deep ocean to the surface. They then are carried worldwide by the conveyor belt and serve to fertilize plankton growth.

But the nearshore Antarctic path back down into the abyss sinks a significant fraction of this recently upwelled phosphate before it can participate in photosynthesis. This shunt is not thought to operate in glacial times (the deep phosphate concentration drops then), presumably because more extensive pack ice seals over the polynyas to close off the shunt.

Making the currently shunted phosphate available to the ocean surface globally, so the story goes, would boost worldwide ocean productivity, drawing down CO2 in the same way as it did at the end of the last warm interval in the ice ages.

My speculation is that switching off this shunt pathway could be done by anchoring balloons to ice upwind of an offshore polynya, so the offshore wind drags it into the opening and closes it. For a coastal polynya[6] where there is no upwind ice to drag, a high shoreline berm or sea wall might be constructed to deflect the valley downslope wind from the nearshore ice edge, allowing ice to form near the shore and anchor itself on a frozen bottom.

Unfortunately, given those thousand-year ocean circulation paths, there is little reason to think that this would be quick enough for a climate fix—though it is possible that the increased fertilization of the Southern Oceans might produce quicker results. Modifying the existing polynya shunt could, however, be useful for the long-term regulation of climate. (Regulating the plankton plantations and augmenting the northern whirlpools are my other two candidates.)

The fate of most such proposals is, of course, that good reasons arise for not implementing them *as proposed*. My push-pull ocean pumps proposal is meant to serve primarily as an easy-to-remember target that defines the response ballpark by being big, quick, secure, powered by clean sources, and inexpensive enough so that a highly-developed country can implement it on its own continental shelf without endless international conferences. Any alternative scheme must pass those same tests.

Now it is time to distinguish between *reversal, relief, reduction,* and *restoration* outcomes.

Sufficient CO_2 *reduction* should *relieve* heat waves and many of overheating's knock-on consequences such as deluge and drought; it should *reduce* the danger of abrupt climate shifts.

In the case of acidification and the thermal expansion of the oceans, we can talk of *reversal*.

But in many state-dependent systems, history also
matters. Destabilized ice sheets may continue to slide
downhill, increasing sea level long after the temperature
trend is reversed. Many ecosystems exhibit regime shifts and
some may prove difficult to *restore*; once the Amazon's rain
forests are destroyed, its unusual regenerative rainfall cycle
will collapse, and therefore plant succession may stay stuck
for a very long time at the stage of fire-prone grasslands,
with little biomass accumulation to counter that which was
lost to fire and rot of the present rain forest.

9
Countering a Methane Burp

The plankton plantations might also prove relevant for countering the effects of methane, so let me outline the various methane threats as they are currently understood.

By now you have probably heard about one or more of the following;

- Natural gas is mostly methane. Gas pipelines leak 1-4% of what they carry. Neighborhood pipes and gas meter seals leak as well. Every time a gas furnace or hot water heater cycles off, some of its dead space methane is vented unburned. (Turning the flame up and down, in the manner of gas stoves, would make more sense than the current use of on-or-off cycling to adjust heat production.)

- Grazing animals have extra stomachs for the slow digestion of grass, and so they burp methane, CH_4, as well as CO_2. That much is "natural." What is anthropogenic has to do with the number of animals. The more grazing animals, the more methane production.

- Methane is what causes mine explosions and, more commonly, the asphyxiation of coal miners. Much makes it into the atmosphere unburned. A century ago, "coal gas" was piped around cities for lighting purposes.

- The frozen Arctic tundra will, as it thaws, produce a lot of methane and release it into the air. There is already a steady leakage of methane from flooded permafrost deposits on Siberia's continental shelf[1]. As the Arctic Ocean warms, there will be more.

- Besides such sources of steady leakage of methane into the air, there can be large releases of methane from the sea floor as methane hydrates break loose from the continental slope during an earthquake. They float to the sea surface and melt, delivering the methane to the atmosphere.

First, the methane basics[2]. Molecule for molecule, CH_4 is about a hundred times more potent as an atmospheric heat-trapping gas than is a molecule of CO_2 on the time scale of decades. The amount of methane in the air has doubled in the last fifty years. Methane now produces 14 percent of the total greenhouse gas contribution to global warming, almost as much as deforestation's 18 percent.

The "natural gas" that comes up a well is mostly methane, with variable amounts of water vapor and CO_2 that must be removed before it will burn reliably.

When organic carbon is digested by bacteria, both CH_4 and CO_2 are released. But the proportion depends on the availability of oxygen. When little O_2 is available (hypoxic and anoxic conditions), little CO_2 can be formed—and so decomposition proceeds more slowly via the methane route.

If there is a lot of O_2 available, decomposition is quicker and it is mostly CO_2 that is produced. There are additional decomposition routes as well; the primary treatment of sewage involves bubbling a lot of air (20% O_2) through the raw sewage, trying to produce a lot of CO_2 before the other decomposition routes can make gases with offensive smells. (Both CO_2 and CH_4 are odorless.)

Let us now imagine that some brief event—something like the 2010 Gulf of Mexico disaster where it took six months to stop the release of crude oil and natural gas from a deep drilling blowout—were to double the amount of methane in the atmosphere. How long would this excess endure?

Much of the new methane from the six-month blip would be removed in ten years by natural processes. (In contrast, a CO_2 blip takes about ten times longer to fall as far—and maybe 25% would still be around a thousand years later.)

A big blip could get us into serious problems, especially via triggering abrupt climate shifts. And a big sustained leak is the leading candidate for some of the mass extinctions in the fossil record[3].

Where might a methane leak come from? Most obviously, from the legacy of our addiction to petroleum: many, many wells have been drilled into depths, looking for oil or natural gas. Most have been abandoned as unpromising, the pipe

being capped at the surface or on the ocean floor. South of
Houston and New Orleans in the Gulf of Mexico, 7,000 wells
were promising enough to construct a platform on stilts,
though 3,000 of them have since been abandoned. The
length of the pipes connecting them to shore would, laid end
to end, reach around the world.

There is also a fair amount of methane found in
association with coal. It is what kills thousands of miners
each year via explosions or asphyxiation (the methane
dilutes the oxygen content of the air so much that you cannot
breathe fast enough to get the minimum amount of oxygen).
The slower seepage of methane mostly escapes unburned
into the air. And since there are a lot of abandoned mine
shafts and scalped mountain tops, it is small wonder that the
highest concentrations of methane in the United States are
found over states with a lot of mining, not over the states
with a lot of livestock that burp methane.

Strip mines and mountain-top removals, while
considerably safer for the coal workers, create many new
paths for venting methane into the air. While we can imagine
sealing up old mine shafts to reduce the rate of methane
leakage, that's not going to work for such grand earth-
removal operations that bleed methane everywhere.

Now we know enough to talk about what a big methane burp
would do to change the climate. Initially it acts like even
more CO2 excess: it overheats the planet and sets us up for

abrupt climate shifts. But we could counter its overheating effects by reducing the atmospheric CO2 accumulation even more.

We probably would not need to build a methane-capture facility on the scale of the plankton plantations. Indeed, we could use plankton plantations to counter the methane's heating effects by reducing the CO2 heating effects to below the 280 ppm baseline. At least in the short run, the methane treatment would be the CO2 treatment, intensified. That is to say, reducing heating via carbon sequestration or reflecting away sunlight would serve to counter methane's overheating effects.

The imperative is to have standby cooling capacity in case a giant burp of methane suddenly makes it into the air. To my mind, we would need plankton plantations even if we didn't have an excess CO2 problem. We need them (or an equivalent), just sitting around with most of the windmills and wave-powered pumps turned off, ready to spring into action to counteract a methane burp.

The two big sources for a methane mega-burp—something far larger than what any malfunctioning well might contribute— are thawing tundra (containing frozen soil from ancient dust storms) and unstable continental slopes (where the continental shelf drops off to much deeper bottoms) in the Arctic and sub-Arctic latitudes.

Thawing allows the tundra's organic matter to be decomposed, producing CO2 and CH4. The Arctic Ocean, formerly roofed over by reflective ice and snow for most of the year, is now increasingly uncovered in the summertime, the dark surface waters absorbing sunlight around the clock. That warms the air above the surface. The winds then warm the surrounding land—which is where the tundra is. Melt ponds spring up like pox marks on a face. Once thawed, the buried organic matter can be decomposed by the usual bacteria—but the buried oxygen quickly runs out, and so methane accumulates, percolating upward.

When the ice capping an Arctic pond starts to melt, eight months of trapped methane escapes into the air[4]. Tundra scientists amuse themselves by drilling a small fishing hole through the ice covering a pond. Enough methane comes whistling out to be set on fire, producing a fiery fountain.

The water in the pond, being denser than the underlying frozen tundra, cuts into the pond's bottom like a wedge, speeding up the thaw and methane production.

Offshore, earthquakes under the continental slope cause avalanches, allowing buried ice, with methane trapped in between the H_2O molecules in their icy matrix (called a methane hydrate or clathrate), to pop loose from the sea floor and rise to the sea surface. And to melt there, releasing the methane directly into the atmosphere.

This is already happening in the North Pacific Ocean offshore of British Columbia and in the Arctic Ocean offshore of Siberia. Fishermen amuse themselves by holding up a piece of ice and setting fire to it.

An excess of earthquakes is expected offshore because sea level is rising. The continental slopes are now under added hydrostatic pressure from above, and where the hillside is unstable, avalanches will occur (they are the common cause of tsunamis). Some will let loose large amounts of "The ice that burns" to float to the surface and melt.

When you know that there are going to be more fires, you stockpile water for firefighting. We now know that methane burps are a serious global threat and so we must provide cooling capacity, held in a stand-by reserve. After the twenty-year cleanup of CO_2, the plankton plantation might serve nicely.

If you have read this far, you are now well-informed about the known climate threats, far better than your neighbors.

Consider sending your favorite chapter to a local newspaper editor as a suggestion for a feature article or editorial. Or post it to a blog to start a discussion. Send it to your lawmakers.

PDFs of individual chapters can be found at *WilliamCalvin.org/media*. Because of what I said on the copyright page, you don't need to seek permission for copying unless you are a book publisher. Just do it.

10
What Will the Greek Chorus Say?

In the drama of the ancient Greeks 2,500 years ago, a tragedy adhered to a certain form. The protagonist was always a good, respected man of great power, often a hero, who acted out of good intentions. Despite this, he had a tragic flaw, or made a tragic error, that put him on the road to catastrophe. He would eventually recognize his problem— but, at least in those classic tragedies, it was always too late to save him from the consequences of his actions[1].

The Greek Chorus stood around on stage, commenting on the dramatic action. Not only did they fill in the background story, but they had a point of view, often setting up an ethical framework by which the protagonist's action should be judged. The chorus helped to evoke the visionary experience that was "the very essence of tragedy."

So what might a Greek chorus have to say someday about a looming climate tragedy?

(Scene: What appears to be the stage of an amphitheater, now washed by waves from the rising sea level. The backdrop shows a sandy beach and a full moon low in the sky, seen through sunset haze. Most actors except the politician are scantily clad, hoping to catch some cool breeze from off the ocean for evaporative cooling. If desired, the politician can be

*shown with an umbilical hose piping in cool air to a
pants leg from his vehicle at stage left. Vacuum-cleaner
hose will suffice. It can be run up the actor's back so that
his bouffant hairdo is made to stand up.)*

Chorus (Prologue, chanted while coming onstage): Seven
years of deluge here, drought there, hunger everywhere.
Coastal cities hit by big waves and bigger hurricanes,
New Orleans writ large.

And everywhere this heat wave, summer after
summer. The death toll is high among infants and senior
citizens. Our air conditioners cause the electrical grid to
collapse. Then even our electric fans do not work.
Buildings are uninhabitable if their windows cannot be
opened. And even evaporative cooling doesn't work like it
used to, because the humidity has become so high from
global warming.

Chorus Leader: Who dares not stir by day must walk by
night[2]. Ghostly pale, we gather at the sea shore each
evening, hoping for a breeze.

Water supplies frequently dry up, causing some cities
to be abandoned. But because things are not much better
elsewhere, half of the fleeing people die. Our neighboring
countries are starving. Their militias raid our border
states for food.

The people are bewildered and angry, half-crazy with
suppressed rage from decades of being worried and

challenged, repeatedly bitten from without and within[3]. The times are desperate.

It feels as if civilization were about to collapse. And without civilized behaviors, the life of man is solitary, poor, nasty, brutish, and short[4] *(Pause, then repeat the last sentence.)*

Chorus Leader (turning to actors): Rumors fly around that it is now too late to fix our climate—that we are already doomed, stuck on the road to ruin, spiraling up into the hothouse. Is that true?

(Three reporters wearing suits come alive as the Congressman begins to speak, take out their pens and raise their videocams.)

Congressman Deny Delay: Of course not. Let's have no more of your unpatriotic climate nonsense—we've proved it's just another spell of bad weather. To get us through it, we're building more coal plants to run the extra air conditioners.

(Reporters sit.)

Physician (rolls eyes up): Just build more coal-fired plants, indeed! You apparently seem to think that fanning the flames is still a good idea. Don't you ever learn? *(Congressman acts as if he heard nothing.)*

ClimateGuy (looking at *Chorus Leader*): To answer your question about inevitability, about ten years ago I would have said no, that we could fix things. I thought then that, politics aside, that we were capable of repairing the climate, that it was both technologically and economically feasible.

But we've become much weaker economically since then. And we've added so much more to the CO_2 burden. Now I'm not so sure that we can get our act together, not as confident that we can clean up the CO_2 in time to save ourselves. For all I know, we're on the slippery slope already.

EconomicsGuy (turning toward *Congressman*): I hope you took in the implications of what he just said. That except for your brand of politics, we could have avoided this mess we're in.

Do you live in your own little world, where cause and effect somehow functions differently than it does for the rest of us? Where two plus two equals five? Where wishes somehow trump reality?

And it isn't that you were blind. Otherwise you and your buddies wouldn't have been so proactive for decades, trying to keep people confused.

Congressman Deny Delay: I resent that, sir! Let me remind you that we are duly elected and doing the will of the people. You lost the election, sir! Don't get in our way! *(Exits stage left, purposefully, followed by reporters.)*

EconomicsGuy (looking stage right and then turning to the others): I've known that guy for years and he didn't used to be like that. He was a good man, even if I didn't like the way he voted. You could speak frankly to him and he would listen. But then he got into the climate skeptic game for his fifteen minutes of fame. And that sort of thing spreads like a contagious disease, winning them elections.

Physician: When something interferes with our self-image, or when we are afraid, guilty, or confused, we tend to deny it. It only takes one of them to trigger denial. Climate change can invoke all four of them.

Chorus: Science, driven by a quest for truth, used to be the common touchstone of what is real and verifiable. But then the fans of business-as-usual spent millions on advertising, to make sure that people stayed confused about the causes of global warming and the consequences. Those manipulators made it seem reasonable to wait before acting.

Next they tried cutting back on funding for climate and alternative energy research.

When that failed to stop the bad news, they tried intimidating the climate scientists, calling for congressional investigations of those who spoke out. The deniers spammed the scientists with lawsuits, forcing the scientists to pay lawyers out of pocket and then take out second mortgages on their homes.

Now the deniers are even questioning the patriotism of those who keep pointing out that the emperor is naked.

EconomicsGuy (pointing at the departing congressman):
Suppose he's turned into a control freak? That bit about
"We've proved it's just another spell of bad weather"
implied that it's the ruling party's determination that is
relevant, not that of the scientists.

Remember when the White House was caught, just
after the turn of the century, letting an ex-oil-company
lobbyist edit what the government scientists said about
climate?

ClimateGuy: And to think that I once was so naïve as to
believe that the facts spoke for themselves.

Reminds me of when the Vatican thought that Galileo
was impinging on the church's prerogative to say what
the heavens contained—such as the Earth being the
center of the universe, around which other heavenly
bodies rotated.

Physician: But still it moves. However, control issues don't
explain very much. He reminds me more of a delusional
patient.

EconomicsGuy: Delusional? Doesn't that mean he believes
something that is false—despite incontrovertible
evidence to the contrary? But unlike most people who
believe impossible things, no amount of rational
argument will budge him?

Chorus Leader: Alice in Wonderland laughed: "There's no
use trying," she said; "one can't believe impossible

things."

Said the Queen: "I daresay you haven't had much practice. When I was younger, I always did it for half an hour a day. Why, sometimes I've believed as many as six impossible things before breakfast."

Physician: Practice helps. But to qualify as a medically-interesting delusion, the belief also has to be contrary to what almost everybody else believes. Otherwise commonplace superstitions and some venerated religious beliefs might qualify as delusions.

EconomicsGuy: There certainly is a vocal minority that goes along with climate denial—though there is some bullying involved. Remember back in 2010 when they first made climate denial a qualification to run for governor or for Congress—at least if you wanted campaign money?

They will force any politician who shows signs of doing actual hard thinking to walk a plank into a sea of craziness[5].

Physician: They may have gone collectively mad, but it's not as if some antipsychotic medication could fix their problem. This is political pathology and, unlike most delusional behavior in individuals, it can be fatal to others.

Chorus Leader: In ancient Greece, should the ruling party harbor delusions about their own strength or cunning, they were likely to lose a war. The victors would usually slaughter the male half of their population, with the women marched off to be sold as slaves. The threat of that fate helped to keep delusional leaders from gaining power because the more pragmatic would be so wary of them.

Physician: I think that the climate delusions started out as that "see no evil, hear no evil" stance of people who want to avoid the subject.

Avoiders may accept the science, yet don't talk about the climate crisis because their personal ethics would then require them to take action. Actually doing something would conflict with their other priorities in life.

'Just pretend it isn't there and maybe it will go away.' Best recipe for disaster ever invented.

But the scale of it! The deny- and delay-mongers have made disbelief in science into a ruling superstition, perhaps to cater to the anti-government business-as-usual crowd that finances them.

EconomicsGuy: There is a loud and intellectually corrupt segment of public life dedicated to fact-denial[6]. Decades ago, when they first became so bold about climate denial, climate change wasn't yet a taboo subject. Now they've gotten coercive about it because they're so afraid that the current banking system will collapse.

Physician: The deniers don't worry any more about getting caught in their lies, having learned that so many voters simply don't pay enough attention to connect the dots.

Journalists can do it but their bosses often have firm orders to discourage it.

And so the deniers can keep using the climate skepticism of the 1970s over and over because, no matter how often you explain that those questions have long since been answered, the misrepresentation doesn't catch up with the deniers.

Chorus Leader: Many people get their news only from the loudmouths of talk radio. And all the constant repetition of the lies indeed works just like Joseph Goebbels told Adolf Hitler it would. Even if someone watches the evening news, it may be a network whose owner has told the news department to always report climate news as if it were still controversial, mere opinion[7].

EconomicsGuy: Lies just lead to more lies. Now they have to lie to keep the economy from crashing. No one wants to lend money anymore, not even for five years, unless persuaded that business-as-usual will prevail and they will get their money back. So the ruling party sets out to fool them, all in the national interest.

Chorus: You can fool all the people some of the time, and some of the people all the time, but you cannot fool all the people all the time.[8]

Physician: Hey, they're channeling Abe Lincoln this time. But Lincoln left out something important.

The deny-delay crowd gets away with fooling folks because Lincoln's "all the time" assumes we have enough time left for the truth to sink in and for us to throw out the unscrupulous. But it's not happening and there's not enough time left.

ClimateGuy: Now if the ruling party was pragmatic, they would be making a big show of projects to repair the climate, burying all of that excess CO_2 and raising hope that way. But they're pretty stubborn.

[Silent musing.]

Chorus Leader: The Resurrection Project has now engraved most of the how-to manuals from the web onto the inside surfaces of aluminum beverage cans, one page per can. Half of college textbooks are now engraved. Even if books were burned and digital records were lost, metal scavengers would discover the engravings when cutting cans open to flatten them into roof shingles. Some would piece books together and read them, speeding the resurrection of civilization.

But after analyses of recent political history were included, all such engraving was banned by the ruling party as defeatist propaganda.

ClimateGuy: And while all countries contributed to the CO_2 blanket, it's the politics of the U.S. that creates a global bottleneck. If you can't get the country that contributed the most to the CO_2 accumulation to start moving, then other countries are going to have trouble asking their own people to make sacrifices.

The U.S. used to have the reputation, left over from World War II and our brief Space Age, of being the can-do innovator that worked around problems. Just the kind of country that you'd expect to take the lead in solving the climate problem.

Chorus: If flying-saucer creatures or angels or whatever were to come here in a hundred years, say, and find us gone like the dinosaurs, what might be a good message for humanity to leave for them on the walls of the Grand Canyon? *(Unrolling banner) WE PROBABLY COULD HAVE SAVED OURSELVES, BUT WERE TOO DAMNED LAZY TO TRY VERY HARD.—Kurt Vonnegut* (1991)[9]

[Silent musing.]

Chorus Leader: This nation was built by men who took risks: pioneers who were not afraid of the wilderness, business men who were not afraid of failure, scientists who were not afraid of the truth, thinkers who were not afraid of progress, dreamers who were not afraid of action.[10]

EconomicsGuy: Right. And now we have leaders who stick their heads in the sand—and are proud of it. Now the rest of the world sees the United States as having gone collectively mad.

They remember the precedents. We prefer not to recall them, but foreign leaders likely remember those mass suicides of unbelievably gullible middle-class Americans. First there was the Jonestown Massacre in 1978, the mass suicide of 900 members of an American cult led by Jim Jones.

Then in 1989 near San Diego, Heaven's Gate members led by Marshall Applewhite—believing that the earth was about to be wiped clean and refurbished—committed mass suicide on the theory that they could thereby escape the fate of others via hitching a ride to heaven on the Hale-Bopp comet, passing by at that time in the night sky.

ClimateGuy: It's hard to blame the rest of the world for suspecting that we'll take the suicidal path for climate. For some reason, they resent the fact that we'll drag them along with us.

As an American, I get a lot of hateful looks from people on the streets when I travel abroad to climate science meetings. So I adopted a British accent and, while in London, bought myself a complete outfit of clothing, shoes, sunglasses—even argyle socks.

[Silent musing.]

EconomicsGuy: At least those who continued to insist that the earth is flat never achieved such political power—nor did they threaten the rest of us with a chaotic collapse of civilization. And the disappearance of civilized behaviors.

Physician: Even worse than the suicidal fools are those proactive fans of Armageddon. They attempt mass murder, hoping to trigger the End of the World.

Remember the first one, Aum Shinrikyo, that got written up[11] in a medical journal as if it were an emerging infectious disease? That sect which attracted so many young graduates from Japan's major universities that it was dubbed a "religion for the elite"? Lots of members and a big bank account, with offices in the U.S., Russia, and Japan? They attracted audiences of 15,000 people even in Moscow.

Their leadership's 1995 nerve gas attacks on the Tokyo subway trains were said to be an attempt to provoke a war that would destroy everyone—except, of course, the faithful. The idea was to blame the attacks on the U.S., triggering a world war, which would then lead to Armageddon[12].

EconomicsGuy: Fools, maybe, but technically capable fools— they brewed their own nerve gases. They turned malicious under the leadership of a juvenile bully who later learned charisma and thrived. But his second in command was a cardiologist and the other subway attackers had applied physics degrees.

It looks as if training in logical thinking isn't a reliable defense against delusional thinking.

[Silent musing.]

Chorus Leader: So are humans afflicted with some tragic flaw?

Physician: Such as susceptibility to delusional thoughts?

That might be because the structured thought which we need for planning ahead and for speaking long sentences—and reasoning about them—appear to have been invented rather recently in our evolutionary history[13]. Maybe only 2,000 generations back, and that's not much time for natural selection to get the bugs out. Or at least, not well enough to withstand modern media onslaughts.

But otherwise, the psychological imperfections responsible for our plight seem pretty minor to me[14], just stuff that is unethical in many cultures— such as trolling for suckers. It doesn't take some urge to suicide or mass murder, triggered by the complexities of modern society.

EconomicsGuy: Yet the consequences of playing people for fools, with misrepresentations the mainstay of many a political ad campaign, are very different than in the days of trying to sell the Brooklyn Bridge to tourists.

Even when the deniers are caught lying, it doesn't seem to make any difference in the next election, what with all the media obfuscation they can stir up. Which just encourages them to try it again and again.

[Silent musing.]

ClimateGuy: And to think that we started the climate damage with really low-tech means—just cutting down trees, then mining coal. That's pickaxe-level technology.

But it led to the dry-rot of civilization where, even though various means to back out of the climate crisis were possible, we actively chose to pretend that the climate problem didn't exist.

(Foreground stage lights begin dimming and then flickering. The Chorus briefly chants an exasperated "Not again." The moonlit beach scene backdrop soon becomes the dominant stage lighting, the actors silhouetted.)

ClimateGuy: Let me quickly say, to end on a positive note, that there's still some chance of hauling ourselves off this slippery slope by a crash effort to clean up the CO_2. Unlike when it was first proposed, back when the economy was good, we'll now need four times as much technology as well as a far bigger dose of good luck.

If we make it, this will still be a very impoverished planet to live on. Those millions of species that recently went extinct won't reappear. We already have major ecosystems to repair and the Amazon's rain forests probably won't ever come back.

EconomicsGuy: You know, that interminable opera of Wagner's, Götterdämmerung, is all about the end of the world. But remember what they say about it. "It ain't over until the fat lady sings." And that Brünhilde sure turned out to be pretty long-winded.

[Pause.]

Chorus Leader (repeated several times as they all slowly leave the stage to walk along the "beach" background):

Here, on the level sand,
Between the sea and land,
What shall I build or write
Against the fall of night?
Tell me of runes to grave
That hold the bursting wave,
Or bastions to design
For longer date than mine.[15]

11
The Great Use-It-or-Lose-It Intelligence Test

Most civilizations of the past have proven fragile[1], snuffing out their own candle. Climate change was usually the final blow, once they had made themselves vulnerable via explosions in population and resource consumption.

Thanks to both history and science, we're the first society to understand what's going on, both with resource issues and climate change.

But presently some political parties make one wonder if human intelligence is mature enough to avoid committing collective suicide—even though we're still technologically and economically capable of repairing the rot we have caused in the foundations. The problem is the political willpower to do it quickly enough.

My chapter title comes from a dinner lecture that I gave in Beijing to the people who created the green revolution in agriculture. (You can view the hour-long video on the World Bank's web site or at global-fever.org). In its written version, the pamphlet starts out this way:

> To fit the magnificence of this setting in Beijing's Great Hall of the People, and the honor of giving the Sir John Crawford

Memorial Lecture, it is well to have a subject of suitable proportions.

I have chosen one of global size and urgent time frame: our climate crisis. We only have one future and one global climate–and now it looks as if we only have one chance to rescue our civilization from collapse and prevent a mass extinction of species during the 21st century.

It is easy to appreciate that one more degree of global warming will seriously reduce crop yields in the tropics, but in the words of climate scientist Claudia Tebaldi[2], "It's the extremes, not the averages, that cause the most damage to society and to many ecosystems." Even if you live where the *average* rainfall stays the same, there will still be more extreme weather such as floods and droughts. That they "balance out" will comfort no one.

Unless you have been keeping up with climate science for the past twenty-five years, you likely do not know how serious the matter has become. The notion that we might slowly get into serious trouble by mid-century has been conveyed by the media and understood by at least some political leaders. But that scenario depends on somehow avoiding sudden shifts in climate in the meantime, instant setbacks at a time when we lack maneuvering room....

Preventing the 2° fever is the Great Use-it-or-lose-it Intelligence Test. And we are dealing with the time frame used centuries ago by Edmund Burke when he said, "The public interest requires doing today those things that men of intelligence and goodwill would wish, five or ten years hence, had been done."

We are already in dangerous territory and have to act quickly to avoid triggering widespread catastrophes. The only good analogy is arming for a great war, doing what must be done regardless of cost and convenience.

We dare not wait until we are weakened before undertaking emergency climate repairs. Our ability to avoid a human population crash will be compromised if economies become fragile or if international cooperation is lost via conflicts. A serious jolt—say, a major rearrangement of the winds—could cause catastrophic crop failures and food riots within several years, creating global waves of climate refugees with the attendant famine, pestilence, war, and genocide.

At the beginning of World War II, Franklin D. Roosevelt used the metaphor of a "four alarm fire up the street" that had to be extinguished immediately, whatever the cost. Our need for fast action on climate deterioration requires devoting the resources necessary to radically shorten the developmental cycle for all carbon burial projects.

Acquiescing in a slower approach is, in effect, playing Russian roulette with the climate gun. The climate crisis needs wartime priorities.

***

Acknowledgements

I am grateful to James J. Anderson, David Archer, John Edwards, Steve Emerson, Richard Gammon, Katherine Graubard, Charles Laird, Edward Miles, James Murray, Gary Odell, Gordon Orians, Julian Sachs, Eric Steig, Alan Trimble, Arthur Whiteley, and Dennis Willows for scientific discussions and comments on earlier drafts. For improving the readability of the manuscript, I am particularly grateful to Larry S. Anderson and Peter G. Rockas.

Explaining the Author

Given the topic of this book, normal people (not just the paid obfuscators in the climate denial industry) are sure to ask: "How can a shrink know anything about climate science?"

It's an understandable confusion. While I am a professor emeritus of psychiatry and behavioral sciences, I am not a psychiatrist. Not even an M.D.

However, that's par for the course. More than half of the faculty at most research-oriented medical schools are Ph.D. types—and that's even true of many clinical departments, certainly mine. For twenty years I was always explaining that I wasn't really a neurosurgeon. For the last twenty, it's been "No, I'm not really a psychiatrist."

Most climate scientists reside in departments such as oceanography, atmospheric sciences, geophysics, and the earth sciences. How does a neurophysiologist find any overlap with them? The journey involved an improbable intermediate stop in anthropology—or at least that part of archeology and physical anthropology concerned with human evolution from the great apes.

I started out in physics. My Ph.D. was in Physiology and Biophysics. And that formal education was what made possible my forays into climate science over the last three decades. I could not have read and understood the research literature without that background.

Though I hadn't thought much about respiration and digestion since my Ph.D. qualifying exams (after which brains became my specialty), the issues of climate science brought it all back. Whenever the oceanographers would mention *convection*,

I'd mentally substitute the physiologists' equivalent term, *bulk flow*. When they got off on the carbon cycle, I'd recall the lectures on the digestive tract and ecology that I used to give when I taught Biology 100 for a few years.

Back in the late 1970s when I was wiretapping single nerve cells as an associate professor of neurological surgery, I took a sabbatical year as a visiting professor of neurobiology at the Hebrew University of Jerusalem and started thinking about the emergent properties of whole circuits of cells—one of which is more precise timing, much needed for accurate throwing and delicate hammering. Someone had just shown that, if you gang together four times as many embryonic heart cells, the beat's irregularity was cut in half—a biological example of the mathematicians' Law of Large Numbers. So there was a way to reduce timing jitter without evolving super-precise nerve cells.

Maybe, I speculated, that's why humans need a bigger brain than, say, chimpanzees. A long 1986 report by Jane Goodall told me that Gombe chimps throw in ways that don't require much accuracy. Throwing is mostly used as a threat. They don't bother to "get set" to throw. Or, for that matter, practice their throws.

Then I noticed that a four-fold increase in cerebral cortex doesn't buy you enough improvement in accuracy. From that I inferred that throwing more accurately might involve temporarily assigning, during "get set," a hundred-fold more neurons to tell the muscles when to move.

That's what got me to reading up on the 6 million years (myr) of human evolution since a chimpanzeelike ancestor. It turns out that brain size didn't start to increase until about 2.5 myr ago. That was also about the time that sharp-edged stone tools became much more prominent. And about that time, the ice ages began.

What did those three trends have to do with one another? Coincidence, or was there some cause-and-effect hiding in there? A brain tripling in size over a mere 2.5 myr is an oddity, suggesting a fast track that needed explaining. I love piecing together multidisciplinary stories like that. (Prior examples: *The Cerebral Code, Lingua ex Machina*, my archaeological explanations for the Acheulian handaxe, and my 1975 article on trigeminal neuralgia.)

The big brain puzzle started taking over my life. I had to pay attention to climate change, as it is often the evolutionary driver that leads to a new species. I heard a lot of discussion of punctuated equilibrium, how "progress" is often compressed into brief periods. In between them, change was usually slow.

It turns out that climate change can also be jerky. Ice age climates are cool, dry, windy, and dusty—but the climate can temporarily shift into conditions that are relatively warm and wet. Back in the 1970s, brief warm-ups were thought to happen a few times within each 100,000 year glaciation cycle. But in 1984 came the first of the ice core evidence that I summarized in *The Great Climate Leap*. It wasn't just a few events but more like a few *dozen* events, some flipping back and forth within only a few centuries, many abrupt transitions taking only five years or so.

If the brain enlarged more easily during episodes of climate change (see *A Brain for All Seasons* and *A Brief History of the Mind*), there were certainly a lot of sudden shifts available for pumping it up.

So after 1984, I read everything I could find regarding abrupt climate change—and for reasons that had little to do

with global overheating. By the time of my 1998 cover story for
The Atlantic, "The Great Climate Flip-flop," I began to realize
that triggering abrupt climate jerks via global overheating might
prove to be more serious than the expected heat waves and
droughts. The large community of climate scientists in Seattle
gave me many opportunities to learn what the pros thought; I've
been participating in their annual retreats for years.

Once I discovered how bad the 2003 Greenland melting had
been, I began to drop everything else and focus on how we
might get ourselves out of this mess. I read up on global
warming per se, whereas my earlier interest was mostly
confined to the abrupt episodes. I wrote it all up in 2008 as
Global Fever.

Then I started collecting modern examples of sudden shifts
that were global in scale even though not as dramatic as the
ancient flips. I'd never heard anyone talk about the modern
quick shifts and near misses but there they were, buried in the
literature of oceanography, meteorology, and rain forest
ecology. Jared Diamond's 2005 book *Collapse* got me to
thinking some more about our present civilization's
vulnerability.

It seemed obvious that we were going to have to clean up
the excess CO_2 accumulation in the air, yet few were talking
about it. No one seemed to realize that all of the good ideas for
slowing CO_2 annual emissions were woefully inadequate for the
bigger job of cleaning up the CO_2 accumulation. Examining
their numbers is what led me to considering push-pull pumps
(they are quite common around medical schools) in my search
for solutions that were sufficiently big, quick, and secure.

Almost no one was talking with a sense of urgency, perhaps because they didn't know about the sudden shifts and near misses since 1976, or because they didn't have much insight into the state transitions that cause instability in such nonlinear systems as heart, nerve, and climate.

Surely, I thought, there were a few climate scientists out there who knew all of what I knew (and more)—but I couldn't rely on them having enough experience in translating complicated science for general readers such as policy wonks and legislators. Or being able to spare the time from their heavy load of teaching, research, and evaluating the situation for the IPCC.

That's where *The Great Climate Leap* and *The Great CO2 Cleanup* came from. Unlike most of my previous books, writing the pair was more a matter of civic responsibility than of choice.

W.H.C.
Seattle

Index

An index has been omitted; however, since the full text is available on the web via WilliamCalvin.org, a search engine can substitute.

Chapter Notes

Chapter 1   *What to Do about Climate?*

1   Joshi M, Hawkins E, Sutton E, Lowe J, Frame D (2011) Projections of when temperature change will exceed 2°C above pre-industrial levels. *Nature Climate Change* 1:407-412, doi:10.1038/nclimate1261

2   Schlosser P, Bönisch G, Rhein M, Bayer R (1991) Reduction of deepwater formation in the Greenland Sea during the 1980s: Evidence from tracer data. *Science* 251:1054–1056. www.sciencemag.org/cgi/reprint/251/4997/1054.pdf

3   Rhines PB (2006) Sub-Arctic oceans and global climate. *Weather* 61:109-118. doi: 10.1256/wea.223.05

4   Våge K, et al (2009) Surprising return of deep convection to the subpolar North Atlantic Ocean in winter 2007–2008. *Nature Geoscience* 2:67–72. doi:10.1038/ngeo382

5   Myhrvold NP, Caldeira K (2012) Greenhouse gases, climate change and the transition from coal to low-carbon electricity. *Environ. Res. Lett.* 7:014019 doi:10.1088/1748-9326/7/1/014019.

Chapter 2   *The Emergency CO2 Cleanup*

1   Calvin WH (2012) *The Great Climate Leap*. ClimateBooks.

2   Brewer PG, Friederich G, Peltzer ET, Orr FM (1999). Direct experiments on the ocean disposal of fossil fuel CO2. *Science* 284, 943–945.

3   The 600 GtC doesn't include the additional amounts of excess CO2 that will come from deforestation and tundra thawing. And it doesn't subtract off the half of the excess CO2 that was absorbed over the years by plants on land; it will not automatically come back out into the air

as we sequester CO2, in the manner of the CO2 absorbed by the ocean surface and now buffered as bicarbonates. But 600 GtC suggests the ballpark for the cleanup.

4   Sokolow R, et al (2011) Direct Air Capture of CO2 with Chemicals. American Physical Society report.

5   Organic molecules are compounds formed in living systems. They have carbon atoms arranged in rings and chains. The main types are carbohydrates, lipids, proteins, DNA, and RNA.

6   Strand S, Benford G (2009) Ocean sequestration of crop residue carbon: recycling fossil fuel carbon back to deep sediments. *Environ Sci & Tech* 43:1000-1007.

7   McNichol AP, Aluwihar LI (2007) The power of radiocarbon in biogeochemical studies of the marine carbon cycle: insights from studies of dissolved and particulate organic carbon (DOC and POC). *Chem Rev* 107:443-466.

8   Boyd PW, et al (2007) Mesoscale iron enrichment experiments 1993–2005: synthesis and future directions. *Science* 315:612. *www.sciencemag.org/cgi/content/abstract/315/5812/612*

    Lampitt RS, et al (2008) Ocean fertilization: a potential means of geoengineering? *Phil Trans Roy Soc A* 366:3919–3945. DOI: 10.1098/rsta.2008.0139

9   Lovelock JW, Rapley CG (2007) Ocean pipes could help the Earth to cure itself. *Nature* 449:403. DOI:10.1038/449403a

    Calvin WH (2008) *Global Fever: How to Treat Climate Change.* London and Chicago: University of Chicago Press.

10 Sheehan J, Dunahay T, Benemann J, Roessler P (1996) A Look Back at the U.S. Department of Energy's Aquatic Species Program—Biodiesel from Algae, NREL/TP–580–24190. US DOE Technical Report. *www1.eere.energy.gov/biomass/pdfs/biodiesel_from_algae.pdf*

Chapter 3   *Flying without a Safety Factor*

Chapter 4   *Sketching Out a Climate Fix*

1   Archer D, et al (2009) Atmospheric lifetime of fossil-fuel carbon dioxide. *Ann Rev Earth Planet Sci* 37:117–134.

2   Ornstein L, Aleinov I, Rind D (2009) Irrigated afforestation of the

Sahara and Australian Outback to end global warming. *Climatic Change* 96. DOI:10.1007/s10584-009-9626-y.

Chapter 5   ***Put the Genie Back in the Bottle?***

[1]   Hansen J, et al (2008) Target atmospheric CO2: Where should humanity aim? *Open Atmos Sci J* 2:217–231.

[2]   Read P (2006) Biosphere carbon stock management: addressing the threat of abrupt climate change in the next few decades. *Climatic Change* 87:305–320.

Read P (2009) Reducing CO2 levels—so many ways, so few being taken. *Climatic Change* 97:449-458.

[3]   Lackner KS (2003) A guide to CO2 sequestration. *Science* 300:1677–1678.

[4]   Houghton RA (2007) Balancing the global carbon budget. *Ann Rev Earth Planet Sci* 35:313–47.

[5]   Strand S, Benford G (2009) Ocean sequestration of crop residue carbon: recycling fossil fuel carbon back to deep sediments. *Environ Sci & Tech* 43:1000-1007.

[6]   Calvin WH (2008) Our climate fix must be big and quick. *arXiv*:0810.2275v1. (Estimates for sequestering organic carbon via sinking sewage in oceans.)

[7]   Ornstein L, Aleinov I, Rind D (2009). Irrigated afforestation of the Sahara and Australian Outback to end global warming. *Climatic Change* 97,DOI:10.1007/s10584-009-9626-y.

[8]   Lampitt RS, et al (2008) Ocean fertilization: a potential means of geoengineering? *Phil Trans Roy Soc A* 366:3919–3945. DOI: 10.1098/rsta.2008.0139

[9]   Note that my CO2 sequestration via relocating decomposition into the ocean depths is quite different from industry's carbon capture and storage via bulk liquid CO2 sequestration. That does not draw down the CO2 reservoir unless burning biomass. Nor is cleaning up the excess CO2 comparable to "clean coal" (which is nothing more than an advertising slogan). Thus my phrase, the *great* CO2 cleanup.

Chapter 6   *The CO2 Cache and the Conveyor Belt*

[1]   Falkowski PG, Laws EA, Barber RT, Murray JW (2003) Phytoplankton and their role in primary, new, and export production. In *Ocean Biogeochemistry* (Fasham MJR, ed), ch.4, Springer.

[2]   Oschlies A, Pahlow M, Yool A, Matear RJ (2010) Climate engineering by artificial ocean upwelling: Channelling the sorcerer's apprentice. *Geophys Res Lett* 37, L04701, DOI:10.1029/2009GL041961.

[3]   Børsheim KY (2000) Bacterial production rates and concentrations of organic carbon at the end of the growing season in the Greenland Sea. *Aquat Microb Ecol* 21:115-123.

[4]   Feely RA, Sabine CL, Takahashi T, Wanninkhof R (2001) Uptake and storage of carbon dioxide in the ocean: The global CO2 survey. *Oceanography* 14:18-32.

       Sabine CL, et al (2004) The oceanographic sink for anthropogenic CO2. *Science* 305:367-371. DOI:10.1126/science.1097403

[5]   Sabine CL, et al (2004) The oceanographic sink for anthropogenic CO2. *Science* 305:367-371. DOI:10.1126/science.1097403

[6]   Feely RA, Sabine CL, Takahashi T, Wanninkhof R (2001) Uptake and storage of carbon dioxide in the ocean: The global CO2 survey. *Oceanography* 14:18-32. At *www.tos.org/oceanography/issues/issue_archive/issue_pdfs/14_4/ 14_4_feely_et_al.pdf*.

[7]   Broecker W, et al (2004) Ventilation of the glacial deep Pacific Ocean. *Science* 306:1169–1172. DOI:10.1126/science.1102293

[8]   McNichol AP, Aluwihar LI (2007) The power of radiocarbon in biogeochemical studies of the marine carbon cycle: insights from studies of dissolved and particulate organic carbon (DOC and POC). *Chem Rev* 107:443-466.

[9]   Boyd PW, et al. (2007) Mesoscale iron enrichment experiments 1993–2005: synthesis and future directions. *Science* 315:612. DOI:10.1126/science.1131669

[10]   Lovelock JW, Rapley CG (2007) Ocean pipes could help the Earth to cure itself. *Nature* 449:403.

       Calvin WH (2008) *Global Fever: How to Treat Climate Change.* London and Chicago: University of Chicago Press. *faculty.washington.edu/wcalvin/bk14*

Oschlies A, Pahlow M, Yool A, Matear RJ (2010), Climate engineering by artificial ocean upwelling: Channelling the sorcerer's apprentice. *Geophys Res Lett* 37, L04701, doi:10.1029/2009GL041961

Chapter 7   *Plowing Under a Carbon-fixing Crop*

[1]   Amon RM, Budéus G, Meon B (2003) Dissolved organic carbon distribution and origin in the Nordic Seas: Exchanges with the Arctic Ocean and the North Atlantic. *J Geophys Res* 14: 1-17. *www.agu.org/journals/jc/jc0307/2002JC001594/2002JC001594.pd f*

[2]   This requires the dilution rate calculation used for "chemostats"; *see* Wikipedia.

[3]   Hilborn R (2007) Reinterpreting the state of fisheries and their management. *Ecosystems* 10:1362-1367. DOI:10.1007/s10021-007-9100-5

[4]   Tsunogai S, Watanabe S, Sato T (1999) Is there a continental shelf pump for the absorption of atmospheric CO2? *Tellus* 51B:701–712.

[1]   Bozec Y, Thomas H, Elkalay L, de Baar HJW (2004) The continental shelf pump for CO2 in the North Sea—evidence from summer observation. *Marine Chemistry* 93:131-147. DOI:10.1016/j.marchem.2004.07.006

[2]   Goyet C, Healy R, Ryan J, Kozyr A (2000) *Global Distribution of Total Inorganic Carbon and Total Alkalinity below the Deepest Winter Mixed Layer Depths.* ORNL Technical report NDP-076 at *www.osti.gov/bridge/product.biblio.jsp?osti_id=760546.*

[3]   Børsheim KY (2000) Bacterial production rates and concentrations of organic carbon at the end of the growing season in the Greenland Sea. *Aquat Microb Ecol* 21:115-123. *www.int-res.com/articles/ame/21/a021p115.pdf*

[4]   Martin JH (1990) Glacial-interglacial CO2 change: The iron hypothesis. *Paleoceanography* 5:1-13. (But see Toggweiler et al 2003 below for the update.)

Chapter 8   *Pro and Con*

5 Toggweiler JR, Murnane R, Carson S, Gnanadesikan A, Sarmiento JL (2003) Representation of the carbon cycle in box models and GCMs, 2, Organic pump. *Global Biogeochem Cycles* 17:1027, DOI:10.1029/2001GB001841.

6 Anderson PS (1993) Evidence for an Antarctic winter coastal polynya. *Antarctic Science* 5:221-226.

Chapter 9   *Countering a Methane Burp*

1 Natalia Shakhova, Igor Semiletov, Anatoly Salyuk, Vladimir Yusupov, Denis Kosmach, Örjan Gustafsson (2010) Extensive Methane Venting to the Atmosphere from Sediments of the East Siberian Arctic Shelf. *Science* 327, 1246. DOI:10.1126/science.1182221

2 Calvin WH (2008) *Global Fever: How to Treat Climate Change.* London and Chicago: University of Chicago Press. See Chapter 12, p.158. *faculty.washington.edu/wcalvin/bk14*

3 Peter D. Ward (2007) *Under a Green Sky: global warming, the mass extinctions of the past, and what they mean for our future.* Smithsonian/Collins.

4 Katey Walter et al (2006) Methane bubbling from Siberian thaw lakes as a positive feedback to climate warming. *Nature* 443, at DOI: 10.1038/nature05040.

David Archer (2007) Methane hydrates and anthropogenic climate change. *Biosci. Discuss.* 4 993–1057 and see RealClimate.org/index.php?p=227.

Chapter 10   *What Will the Greek Chorus Say?*

1 A modern example of a classic-style tragedy (though lacking a Greek chorus) is Arthur Miller's second play, *All My Sons* (1947). That the protagonist always understood too late doesn't make Greek tragedies pessimistic. I consider it a playwright's technique for conveying visionary insight in a manner more likely to stick permanently in memory and promote thinking ahead.

2 William Shakespeare (1581) *King John* I.i.172

3 Adapted from a passage in Greg Bear's *Quantico* (2007).

4   Thomas Hobbes (1651) *Leviathan.*

5   Timothy Egan (2011).
    opinionator.blogs.nytimes.com/2011/06/09/the-reluctant-
    mormon/?hp

6   Timothy Egan (2011) "Republicans in a congressional panel
    declared, by a majority vote, that climate change caused by humans
    does not exist. The majority of the House then voted to get rid of
    federal funding for the world's finest scientists in the field to study
    the changing earth, through the Intergovernmental Panel on
    Climate Change. Blink, blink, just like that — our representatives
    wished away the future." Twister's Tale. *New York Times* at
    opinionator.blogs.nytimes.com/2011/05/26/twisters-tale/?hp.

7   *Sent: Tue Dec 08 12:49:51 2009*
    From: Sammon, Bill [Fox News Washington managing editor]
    "Given the controversy over the veracity of climate change data...we
    should refrain from asserting that the planet has warmed (or
    cooled) in any given period without IMMEDIATELY pointing out
    that such theories are based upon data that critics have called into
    question. It is not our place as journalists to assert such notions as
    facts, especially as this debate intensifies."
    See *climateprogress.org/2010/12/15/leaked-email-fox-news-
    sammon-cast-doubt-on-climate-science/.*

8   Abraham Lincoln.

9   Kurt Vonnegut (1990) *Fates Worse than Death: An Autobiographical
    Collage.*

10 American theatre critic Brooks Atkinson.

11 *www.cdc.gov/ncidod/eid/vol5no4/olson.htm*

12 See *en.wikipedia.org/wiki/Aum_Shinrikyo.*

13 Calvin WH (2004) *A Brief History of the Mind* (Oxford University
    Press).

14 Markowitz EM, Shariff AF (2012) Climate change and moral
    judgment. *Nature Climate Change* 2:243-247. DOI
    10.1038/nclimate1378

15 A. E. Housman (1936) *Poems.*

Chapter 11 *The Great Use-It-or-Lose-It Intelligence Test*

[1] Diamond J. (2003) *Collapse: How Societies Choose to Succeed or Fail.* New York: Viking.

[2] Claudia Tebaldi et al, "Going to Extremes," *Climatic Change* (December 2006). See *www.ucar.edu/news/releases/2006/wetterworld.shtml*.

www.ingramcontent.com/pod-product-compliance
Lightning Source LLC
Chambersburg PA
CBHW051327170526
45166CB00002B/717